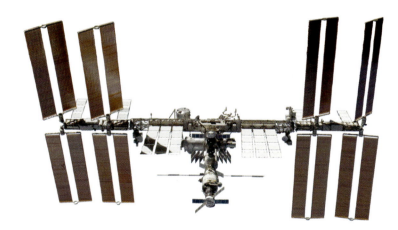

StarTalk

With Neil deGrasse Tyson

Everything You Ever Need to Know About Space Travel,
Sci-Fi, the Human Race, the Universe, and Beyond

NATIONAL GEOGRAPHIC

WASHINGTON, D.C.

To all the fans of *StarTalk,*
whose insatiable appetite for science has inspired
the rich buffet on which this book is based.

CONTENTS

"Our universe is filled with secrets and mysteries, leaving us with many questions to be answered. We find ourselves searching for those answers as the very fabric of space, science, and society are converging. Here, for the first time, these worlds collide!"

—*STARTALK RADIO* INTRODUCTION

Introduction

With those words, and the support of the National Science Foundation (NSF), the first broadcast of *StarTalk Radio* beamed out in 2007 across all of space and time. What has followed—on the radio, on the Internet, and on the National Geographic Channel—is now a movement that has thrown science and pop culture together into a whirling singularity and hurled the result out into our universe for all to enjoy.

Conversations on StarTalk *can delve into viruses, like HIV, and medical advances.*

If you're new to *StarTalk,* you may be wondering: What am I getting into? The millions of fans out there have, no doubt, millions of answers to this very good question. Maybe you can get a sense from real data. In a survey of three arbitrarily chosen *StarTalk* audience members, we've obtained these three assessments:

1. *"StarTalk* is a fun show with an intellectual science bent."
2. *"StarTalk* is a discussion of pop culture within the context of science."
3. *"StarTalk* is a way where you can learn science-y stuff through non-science-y stuff—and it's funny!"

What is *StarTalk*? The name says it all: Stars talk. Movie stars. TV stars. Comedy stars. Stars of all other kinds, as well—artists, activists, authors, opinion shapers, policymakers, and, of course, scientists. Sometimes, the talk is about stars, like the kind that

Astrophysicist Neil deGrasse Tyson hosts the StarTalk radio and television talk shows.

On StarTalk, *Neil and guest stars discuss all matters of the universe.*

have nuclear fusion in their cores and whose light twinkles as it passes through our atmosphere. Mostly, though, it's about anything and everything under those stars; and it's not an interview or a publicity piece—it's a conversation. *StarTalk* has celebrity, and comedy, and crazy egg-headed science. In a world where compartmentalization is increasingly the norm, such seemingly disparate elements, at first glance, would seem hard to combine. *StarTalk* reminds us all that they're not in the least disparate; not only do they play nice together, they've always been peas in a pod—we all just forget that sometimes. On *StarTalk,* science and society link arms and do-si-do in a spirited square dance of laughter and shared discovery. On *StarTalk,* everyone can love art, music, poetry, politics, engineering, science, and math all at the same time. Every pursuit and pursuer—be they scientifically thinking comedians or comedically thinking scientists, or anyone else—come together, inspired by the feeling of awe that comes from pondering the mysterious. Albert Einstein called that feeling "the source of all true art and science," and *StarTalk* is its embodiment—a joyous mind meld of all ways of thinking and being!

Find out what makes an aurora borealis light up the sky, on page 105.

Now you hold in your hands the result of another mind meld: an evolution that happens when digital and print media collide. The skilled team at National Geographic Books has helped *StarTalk* jump from photons to paper, to bottle the lightning of the show's spirit and character into a package you can enjoy even if you've run out of battery power or aren't getting a wireless signal. On these pages—just like on the radio, podcasts, and television—stars talk. They're not "reported on," interviewed, or exposed, like they might be by a news show, magazine, or tabloid. We can see what they said and hear in our heads how they said it. Want to know why Stephen Colbert loves having scientists on his show? How Mayim Bialik feels about the world of *The Big Bang Theory*? Or what Dr. Neil deGrasse Tyson thinks about time travel, visiting Mars, or Superman's reproductive potential? It's all here on these pages, in their own words. On *StarTalk,* there is no "they." There is only "we." We are all invited to partake, unabashed and unafraid, without bias or judgment, in the interchange of ideas, imagination, and humanity's evidence-based understanding of our universe, our world, and ourselves. That's how science works.

Now get busy. Turn the page and discover what *StarTalk* is. Let's do this! And as Neil always says when he bids the audience farewell: Keep looking up. ■

Spacecraft, like Soyuz, are always a hot topic—and StarTalk *gets real insight into them from astronauts.*

Satellites orbiting in space transmit signals—including StarTalk*'s— to satellite dishes on Earth.*

About This Book

nside this book, we've taken the best of *StarTalk* from the airwaves to the page. You'll find the voices of guests who have appeared on the podcast and television show, including actors, comedians, politicians, astronauts, entrepreneurs, millionaires, physicists, neurologists, and biologists, just to name a few. We'll tackle questions that have led some of our greatest discussions on *StarTalk* over the years, and dive deeper into the science behind our universe. And we've got a whole lot of that universe to cover. This book is divided into four sections to make it easier for you to navigate your way through the world of *StarTalk:* "Space," "Planet Earth," "Being Human," and "Futures Imagined." Within these pages you'll discover the mystery behind black holes; real ideas for tackling climate change; just what makes us humans tick; and the amazing science behind science fiction. And there's a lot to take in. Just take a look at what these pages have on offer:

Love what you're reading and want more? We've made it easy for you. Look for the orange bar above the text: That's the name of the *StarTalk* episode this conversation came from. Go to star talkradio.net to hear more. And look to the green text for insightful quotes, or a deeper insight into the topic at hand.

We jump deep into the topics on every page—but some of our best stuff comes from the hundreds of guests who have appeared on *StarTalk* over the years. The orange quotes throughout this book are the wise—and often hilarious—words of these stars. Use the "Did You Know" facts to impress your friends at trivia. The "Think On This" sidebars answer the initial question posed on the page, but from a different angle.

Not all of these spreads come directly from the show—we've created collections in each of the sections in this book to expand your knowledge even more. Look to these pages for timelines of food in space, the history of apocalyptic theories, the fashion sense of sci-fi dramas, and much more.

There are a lot of other informative sidebars throughout this compendium—each with its own flavor. Here's a handy chart of sidebar icons to make it easier for you to navigate:

Laugh Out Loud: This sidebar features some of the most hilarious quotes from our seriously funny guests.

Tour Guide: Learn all you need to know about our universe—from space station bathrooms to who has the best sex.

Back to Basics: You may be scientifically literate, but that doesn't mean you know it all. This box explains harder-to-grasp concepts.

Conversation: One of the best parts of *StarTalk* is the banter that arises in the studio. Look for these witty back-and-forths in our "Conversation" sidebars.

Biography: We've highlighted remarkable scientists who have changed how we view our world.

Drink of the Evening: Neil has developed cocktails with bartenders at the *StarTalk Live!* shows. Get those recipes here.

Neil Tweets: Neil is a prolific tweeter. And we've brought his social media feed to these pages with some of his best tweets.

SPACE

On a bright sunny day or an inky black night,
we all wonder sometimes: What's out there, out of sight?
Beyond Earth's friendly confines, an entire universe awaits
our exploration. But wait! We need to be ready before we
go; we need to know how to get there, how our bodies and
minds will take the journey, and what to expect when we
arrive. Plus, we want to enjoy ourselves, right? Mere survival
isn't the goal; we want to travel in style! After all,
who knows if we'll meet anyone along the way . . .

"Mars is a geologist's dream place, but even if you're just a tourist, it would be stunning, too. You know what I would do? I would not only look down, I would look up, and I would take a picture of Earth in the Martian sky."

—DR. NEIL DEGRASSE TYSON, ASTROPHYSICIST

CHAPTER ONE

What Do I Pack for Mars?

Cooped up in a space capsule for three years? Most of us have a hard time sitting in a car for three hours! That's what we'll have to do, though, to make it to our planetary neighbor, the red planet—Mars.

We'll have to bring a lot with us, too. Sure, Mars may have some of the things we humans need: Vast underground frozen oceans may provide water; mineral resources may provide materials for building and growing things. So what *doesn't* Mars have that we'll need?

Mars isn't our only destination, of course. What do we need for space travel? We're really packing for generations of humans in space, well beyond the familiar confines of low-Earth orbit. Can humans handle living in space? We already have been—astronauts have continuously done so, all this century, in low orbit over Earth, on spaceships such as the International Space Shuttle (ISS). Thanks to their hard work and sacrifice, we've learned a lot—about what it's like to live in space and what we need to take up there with us.

The surface of the red planet is barren and inhospitable to human life—for now.

For decades, astronauts' escape suits have been bright orange for visibility in an emergency.

Do Astronauts Get Taller?

An era of human spaceflight ended on July 21, 2011, when the final space shuttle mission (STS-135, *Atlantis*) and its crew—Capt. Christopher Ferguson, Col. Douglas Hurley, Col. Rex Walheim, and Sandra Hall Magnus—landed at Kennedy Space Center in Florida.

As *Atlantis* touched down, its crew had collectively logged 11 spaceflights and more than 262 days in space. All that combined astronautic experience, though, would still be about one-fourth of that experienced by each human on a mission to Mars.

What happened to their bodies while they were up there? For one thing, they grew a total of about eight inches taller!

▶ **STRETCHING UP** Since gravity isn't tugging downward on their bones, astronauts in space can gradually grow up to 3 percent taller—nearly two inches for your typical spacefarer. Blood also flows more readily out of their feet and up into their heads, giving astronauts that puffy-cheeks look in zero gravity.

▶ **THE DOWNSIDE OF GETTING TALLER** That new-found height comes with a price, though. An astronaut in zero g can lose bone density up to 10 times faster than a 90-year-old person suffering from osteoporosis on Earth. After months of space travel, brittle bones can break from even a minor fall. Muscle strength can drop drastically, too. ■

"You squish down pretty quickly, as soon as gravity hits you. You stand up—whoosh. Gravity is a pretty pervasive force."

—DR. SANDRA HALL MAGNUS, MISSION SPECIALIST, STS-135

Who'd Be the Ideal Candidates for Mars?

E very person on a space mission is precious cargo, and the broader the set of skills and interests the crew spans, the more likely the mission will succeed. Should we send artists and poets into space? We already do! They just also happen to be highly trained scientists, engineers, pilots, and technicians, plus great writers and speakers—even if most are quite modest about it. And someday in the not-so-distant future, space dreamers like musician Josh Groban will be packing for off-world travel, too.

> "When I was a kid, my favorite costume was a space suit. All day long, every day, I'd wear a space suit. And I was not Josh Groban from Earth . . . I would say, 'I'm John from another planet.' "
>
> —JOSH GROBAN, MUSICIAN

What characteristics make the best astronaut—someone who will have to live with the same few people in tiny spaces for very long periods of time? Says Mary Roach, author of *Packing for Mars,* channeling certain cultural stereotypes: "The Japanese are, speaking very generally, good astronauts for a number of reasons: small payload, used to tiny spaces, not much privacy, and also, again, obviously speaking in generalities here, raised to be not confrontational and aggressive but polite and respectful." ∎

Japanese astronaut Kimiya Yui

DID YOU KNOW
Astronaut and musician Chris Hadfield released an album he recorded on the ISS. *Space Sessions: Songs From a Tin Can* hit number 10 on Canadian charts in 2015.

LAUGH OUT LOUD ▶ **With Chuck Nice, Comedian**

"You know, it just occurred to me that with the stress and the food, Jamaicans would make great astronauts. You call them 'Rastanauts.' Just like, 'Ya, mon, everything's irie. Inhale. OK, now eat that, mon. It's delicious.' Everything you need to deal with, your stress and food issues, are taken care of with one little Rasta puff. If your fellow astronaut is on you, 'Ya, mon, go ahead and smoke this.' "

When Are We Going Where Hundreds Have Gone Before?

Spacewise, we humans have been backtracking a little since the 1970s. For more than 40 years we've gone—hundreds of us actually—only as far as low-Earth orbit, barely a hundredth of the distance to the Moon. And now, of course, we're all impatient to head for Mars, hundreds of times farther than the Moon. During the majority of a Mars trip, though, it'll be pretty much like living in low-Earth orbit—on a spaceship, with our minds and bodies dealing with isolation,

"We're learning so much on the space station about how to survive for two and a half years, which is a typical mission template . . . What we're learning today on the space station is going to make those future missions successful."

—STS-135 CREW MEMBER

confinement, radiation, microgravity, and more. So until we have the funding and technical know-how to head for the red planet, the experiences of astronauts on the ISS amount to the most valuable data we have about such a trip.

At the current rate of technological progress, in about 25 years we humans may finally be headed to Mars. If so, then the crew of that first mission probably isn't out of high school yet. So start planning, kids! ∎

U.S. Astronaut Bruce McCandless gets a stunning view on a space walk.

A space suit from Project Mercury

A suit worn on an Apollo mission

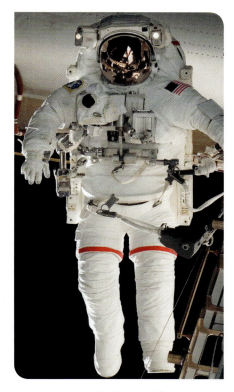

An ISS extravehicular mobility suit

What Should I Wear to Go Exploring the Solar System?

"Stay here on Earth. You can live the longest on Earth."

—DR. NEIL DEGRASSE TYSON

But, hey, if you're going to disregard Neil's advice and explore the solar system, you need to stay alive and well—and you want to look good doing it. NASA's Solar System Exploration Research Virtual Institute (SSERVI)—dedicated to exploring Earth's moon, near-Earth asteroids, and Mars's two moons—can help.

Each of these destinations will have its own challenges. Most asteroids, for example, have so little gravitational pull that you can launch off the surface into space with just a running start. So your space suit will need a way to keep you connected to the ground! You'll also be vulnerable to plasma discharges, so your space suit will need enough metal to shield you from asteroidal lightning. ■

A space-suit glove allows for mobility.

Nichelle Nichols as Lieutenant Uhura in Star Trek

How Did Nichelle Nichols Change the Face of NASA?

The starship Enterprise

Nichelle Nichols broke the television color barrier as *Star Trek*'s Lieutenant Uhura, the African-American communications officer on the starship *Enterprise.* When Nichols was asked by NASA to be part of the publicity campaign to recruit astronauts, her insistence that the astronaut corps include members of all races and genders helped create the diversity of humans in space today. As she puts it: "I said, 'I will help you, but I'm going to get you the most qualified people of anybody you've ever had, and therefore I don't want to hear any excuses, because they're going to apply in droves . . . and if there's not one woman and one person of color . . . I'll be your worst nightmare.'"

Little-known fact: Uhura's first name was never revealed in the *Star Trek* TV series. In the 2009 franchise-rebooting *Star Trek* movie, Spock appears to call her Nyota, but when Kirk tries to make it official, Spock says: "I have no comment on the matter . . ." ■

Packing for Mars (Part 1)

You Made Radiation Shields out of WHAT?

Creativity always makes a difference—sometimes, it can even be a matter of life and death. Here's an example of truly thinking out of the box.

To travel and live in space, we need to protect ourselves from hazards unique to that environment. We also need to attend to our own sanitary issues. So let's get creative and do both with eco-friendly, readily available materials.

"NASA has a device," explains Mary Roach, author of *Packing for Mars.* "It's kind of like an Easy-Bake Oven, where you would take [human or animal] waste material and kind of plasticize it into a tile . . . It's a good radiation shield."

Say whaaaa? Yes, you heard right. It's simple, actually. Certain types of harmful radiation that are likely to hit Mars-bound spacecraft can't be stopped by lead or other metal shields. Layers of matter rich in hydrogen would protect the astronauts much more effectively. Such material is generally rare in space, but fecal matter fits the description—and a healthy adult produces about 100 pounds of poop per year. So why not process astronaut waste into tiles or bricks and line a spacecraft with them? It's reduce/reuse/recycle at its best. ■

On a Mars colony, there are millions of miles of land and billions of gallons of frozen water available to recover from a catastrophe. On a spaceship, one major disaster and it's adios, muchachos— mission's over.

TOUR GUIDE

Who's Afraid of a Micrometeoroid?

Micrometeoroid hits are rare, because there is very little natural solid material naturally in space. But when they happen, look out. Just listen to astronaut Sandra Hall Magnus: "Something as small as a dust particle can do some damage at 2,500 miles per hour. That's typically what we see, those small little hits. If you get something the size of a dime or a quarter, you're in trouble." Just as on Earth, we humans create plenty of environmental hazards in space. In 2007 the Chinese government intentionally destroyed an old weather satellite to test a missile system; the impact created thousands of pieces of debris that will stay in low-Earth orbit for decades.

THINK ON THIS ▶ Can You Get Cabin Fever in Space?

As astronaut Mike Massimino puts it: "A lot of it is psychological, the cabin fever." It's a lot like being stuck in an isolated mountain hotel, or so comedian Chuck Nice thinks, asking: "You ever have the urge to just break through with an ax and go, 'Here's Johnny!'?" Answers Massimino: "No. And that's why we don't have axes." On a spaceship, safety comes first.

Are We Floating in a Tin Can?

Scientists and engineers have been imagining and designing space stations—real ones, not crazy fictional Ringworlds and Death Stars—for more than a century. Working models, however, have only been around for about half that time.

||||||||||
◀ SKYLAB 1973

The United States' first space station was almost destroyed during its launch, but the crew saved it in the world's first major in-space repair. Three human crews in all visited the 86-foot-long station, but delays to the space shuttle program doomed Skylab to fall back to Earth in 1979.

||||||||||||
▶ SALYUT 1 1971

The first space station occupied by humans launched on April 19, 1971. The three Soviet cosmonauts orbited for 23 days, but, tragically, suffocated on the return to Earth when their ship sprang a leak. Salyut 1 was de-orbited six months after its launch.

IIIIIIIIIIIII

▶ MIR 1986

The U.S.S.R. launched a station similar in size to Skylab in 1986. The station—aptly named Mir, or "Peace"—survived the fall of the Soviet Union. Mir hosted 125 humans during 15 years of operation before being de-orbited on March 23, 2001.

IIIIIIIIIIII

◀ TIANGONG 1 2011

The first Chinese-made space station held two crews of human visitors for several days in 2012 and 2013. It was the first of a series of stations, expected to culminate in a large ISS-like modular space station that will be launched in the 2020s.

IIIIIIIIIIII

▶ INTERNATIONAL SPACE STATION (ISS) 1998

The ISS emerged from growing world cooperation and shrinking space budgets. Five space agencies jointly run the station, which has been continuously inhabited since November 2000 by spacefarers—including the world's first space tourists.

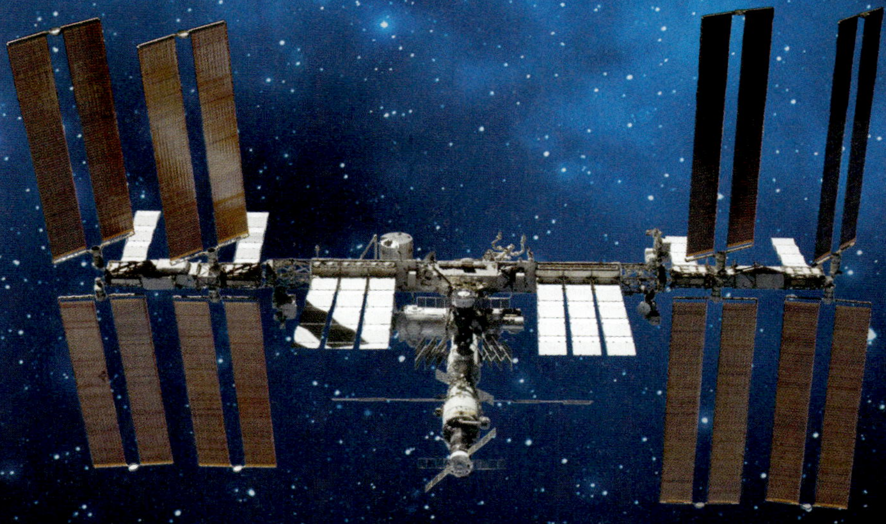

Scientists simulate Mars conditions in Utah.

What Can Antarctica Teach Us About Mars?

An Antarctic expedition reflected in a monument at the South Pole

You get to Mars and what you see is a vast, inhospitable frozen desert. So, coincidentally, is Antarctica. With its five million square miles of brutal isolation, it's the perfect place to study the effects of extended space missions on human health and psychology.

At Concordia, a European science station based in Antarctica, about a dozen intrepid people spend months at a time together in perpetual darkness, farther from civilization than the International Space Station is from Earth.

What happens to a person in that sort of situation? Here's what Mary Roach, author of *Packing for Mars,* says: "After about six weeks, you get this irrational antagonism that sets in, where the very things you loved about your crewmate initially begin to just drive you crazy."

It's not only your mood or your attitude; it's your body. Your immune system weakens. Your hormone levels go crazy. You can't sleep; you can't eat—it's like having perpetual jet lag. Figuring out ways to overcome these challenges will be essential to a successful Mars mission. ■

"There's a story circulating that some space-shuttle astronauts tested 10 different sex positions. NASA disavows all knowledge of such."

—DR. NEIL DEGRASSE TYSON,
ASTRO-NOT-TOUCHIN'-THAT-ONE-IST

Packing for Mars (Part 2)

Is Sex in Space the REAL Final Frontier?

O K, parents, make sure your children aren't reading over your shoulder. Regular social interaction includes, well, some adult-oriented activities.

In Arthur C. Clarke's 1993 novel *The Hammer of God,* it is suggested that Mars has just the right level of gravity to make human copulation maximally enjoyable. No scientific studies have been conducted to test that hypothesis—yet. In the microgravity of the Mir Space Station, though, we may have had . . . umm . . . some preliminary comparison data.

"I asked some cosmonauts about this," Mary Roach, author of *Packing for Mars,* says. "Cosmonauts are fairly forthcoming on things, particularly after a shot of whiskey . . . He said, 'Yes, Mary, people ask me this all the time. They are saying, "Sasha" '—his nickname was Sasha—' "Sasha, how are you making sex in space?" ' And he goes, 'Of course, by hand.' "

There's another private activity in space, even less romantic, that may be even more of a new frontier. Ever heard of the "positional trainer"? Astronauts use it for learning how to relieve themselves in microgravity. There's a camera below the rim, and the user watches on a TV monitor to know when the angle is just right. As Roach says: "In zero gravity, you don't sit on a toilet; you hover." ∎

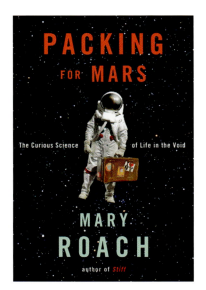

Roach's best-selling book, published in 2010

TOUR GUIDE

Neil's Tips on Martian Sightseeing

"I think of Mars as the American Southwest. There are beautiful structures that look great in the setting sun. There are different terrains that attract different kinds of interests."

"Mars has the largest volcano in the solar system, Olympus Mons."

"If you're into valleys and riverbeds, you'd go to Valles Marineris [Mariner Valley]."

"You can go to floodplains, where water once flowed. And you have river deltas."

What Would It Take to Get Neil to Go to Mars?

As the possibility of moving to Mars comes closer to reality, some people are jumping at the chance and others are saying, "No way." What about Neil? He would take on the challenge as long as it could be a family adventure.

"If I could bring my family, and get a good [online media subscription service unnamed!] account and some books," Neil, the astro-settler-ist, proposes. "My wife is very well educated. We could both totally space-school—homeschool, spaceship-school—our kids, so then it's a family trip. I could totally do that."

Like all early pioneers who move beyond the

"Please, teach us again how great Earth was."

—JOHN OLIVER, ACTOR AND COMEDIAN

frontier into the new and unknown, Neil would pull up his earthly stakes, pack everyone into the Martian covered wagon, and blast off together. But comedian Eugene Mirman has a good point: "Yeah, but then your kids would be furious!"

Ah, not everyone sees migration to Mars as such a *Little House on the Prairie* utopian fantasy. The romanticization of the pioneer experience—immigration from Europe to the Americas, for example, or the settling of the western United States—glosses over the hardships, heartaches, and extreme danger along the way and after arrival, especially for children. ∎

Neil discusses a future colony on Mars in Cosmos—*could he be a part of it?*

Moonwalker and pro-Mars activist
Buzz Aldrin at Stonehenge in the U.K.

How Would Buzz Aldrin Colonize Mars?

Water-ice clouds hover in the Martian atmosphere.

Dr. Buzz Aldrin has written extensively—both fiction and nonfiction—about space travel, exploration, and colonization, including the book *Mission to Mars* (2013). He has proposed a plan, using a system of spacecraft continuously moving back and forth on orbits around Earth and Mars, to get people and necessities to Mars and establish a colony in a 20- to 30-year time frame. First, he believes a base on Phobos (the larger of Mars's two moons) should be built robotically. With enough commitment of resources, he argues, these plans are all technically feasible.

Maybe more important, *why* does Buzz want us to colonize Mars? For the same reason he went to the moon in 1969. It will be another giant leap for mankind. ■

"They will be the most remembered, the most talked about [people] that have ever set foot on Earth. Because they pioneered something that nobody ever did, and they carried it out."

—DR. BUZZ ALDRIN, ASTRONAUT

Astronauts train in a pool with a model of their Apollo spacecraft in 1966.

Why Did You Become an Astronaut?

"There is something about doing something borderline insane that does actually inspire people. Going to the moon was crazy—that was what was so incredible about it. Going around Mars barely makes any sense—so it's inherently inspirational . . . That's what America is built on. It's built on doing something that makes almost no sense to anyone else at the time."

—JOHN OLIVER,
ACTOR AND COMEDIAN

Neil Armstrong and Buzz Aldrin were military pilots before becoming astronauts. Col. Chris Hadfield and Dr. Mike Massimino, two later-generation astronauts, took different paths.

Col. Hadfield decided to become an astronaut when he was a boy: "I looked to my heroes Neil and Buzz and Mike Collins . . . Astronauts fly, so I learned to fly. The odds of being an astronaut are so tiny, that I wanted to do something that would lead to other things that sounded interesting to me anyway."

"I dreamt about being an astronaut . . . watching Neil Armstrong and Buzz Aldrin walk on the Moon," says Dr. Massimino. "Once I got out of college, I decided that's what I wanted to try to do. Went to graduate school, got my Ph.D., and started applying while . . . in grad school."

Astronauts all go through years of physical and psychological training. But the desire to go out into space, and the way to get to that threshold is what makes the corps of spacefarers diverse, capable, and strong. ■

What's It Like on the International Space Station?

"**W**hen you first get into space, it feels a bit odd. It is a very unusual situation to be in. But depending on the person, your inner ear settles down very quickly. Mine settled down right away, and I had very little trouble adjusting to zero g."

Astronaut Dr. Shannon Walker shares her experience while aboard the ISS, Expedition 25: "There are so many things you can do when there's no gravity around. Of course, a lot of the fun things to do are things like playing with your food and making little bubbles and having them go here and there . . . Of course, we have to clean everything up, so we have to temper our excitement with the realities of keeping the station clean.

"I do look out the window a lot, whenever we have a chance to. The ground team keeps us pretty busy, and so we don't get to spend many hours looking out the window." ■

Astronauts in zero g aboard the ISS

TOUR GUIDE

How Do You Use the Bathroom in Space?

In space, even the mundane is an adventure. During the Apollo era, astronauts relieved themselves into plastic bags cupped over their crotches or taped to their buttocks. Now, there is a suction-based toilet that pulls the waste away. On the ISS, the liquid waste is purified into drinking water; solid waste is stored in containers until there's enough to launch it out of the station to burn up in Earth's atmosphere. (Maybe that shooting star you saw last night was actually a fiery fecal payload!)

"Your angle of approach on a space toilet is important . . . It's a docking maneuver."

—MARY ROACH,
AUTHOR OF *PACKING FOR MARS*

What Did Apollo 11 Mean to Mankind?

I f your family members kept one newspaper clipping from the 20th century, it's probably the iconic *New York Times* front page from July 21, 1969, with the headline "Men Walk on Moon." One day earlier, an estimated 500 million people around the world had watched Neil Armstrong take that "one small step for a man, one giant leap for mankind." The news even made the front page of *Pravda*, the main newspaper of the United States' Cold War enemy, the Soviet Union—albeit with a smallish front-page bit, continued on page 5. No matter where we travel in the future, that first Moon landing will always be a milestone of history: the first time humans walked on a world not their own. ■

"Five hundred years from now, it's the only thing they're going to remember about the 20th century."

—COL. R. WALTER CUNNINGHAM, LUNAR MODULE PILOT, APOLLO 7

"[The] most indelible memory was approaching the Moon and flying through the Moon's shadow, so that the Moon was eclipsing the sun and we could see the corona all around the Moon . . . we could see the craters and the valleys and the plains in a blue, gray three-dimensional view that was spectacular . . . remarkable, but imperceptible to a camera. But to the human eye was wonderful."

—NEIL ARMSTRONG, COMMANDER OF APOLLO 11

"My family pulled the dining room table into the living room, so that we could watch the landing, and Armstrong walking on the Moon. It was the first time I ever saw my father, who was an immigrant from southern Italy . . . cry."

—DR. CAROLYN PORCO, PLANETARY SCIENTIST

*" 'Contact light. Engine stop.' We got there . . .
That was really opening the door to every piece
of exploration to ever follow that. Without having
done that, you couldn't do the rest of it."*

—DR. BUZZ ALDRIN, LUNAR MODULE PILOT OF APOLLO 11

*"I was at the launch. So nothing will beat seeing
the three guys walk by me, coming out of
the operations building, early in the morning,
on their way to the Moon. That's like watching
Columbus sail out of the harbor."*

—DR. JOHN LOGSDON, FOUNDER OF SPACE POLICY INSTITUTE,
GEORGE WASHINGTON UNIVERSITY

*"I remember very vividly the landing,
and the fact that they were down,
almost out of fuel, when they finally set
down. It was getting very, very tight,
and we didn't know whether they were
really going to be able to land or
whether they were going
to have to abort out of there."*

—SENATOR JOHN GLENN,
ASTRONAUT ON FRIENDSHIP 7 AND STS-95

*"As Neil [Armstrong] set foot on the Moon,
I was sitting on the hood of a car, listening to the
broadcast on the car radio. There was a girl beside
me, and what was taking place on the Moon was
not my first priority . . . I will say this: I remember
very well what was taking place on the Moon;
I remember very little of what was taking place
on the hood of the car."*

—DR. ROGER LAUNIUS, FORMER CHIEF HISTORIAN OF NASA

"A total round-trip Mars mission is going to last three or four years, so what you really want is gardens on Mars. Create a hab module, and you can grow pigs and cows if you're carnivorous, or celery and carrots if you're veggie, and, you know, go to town."

—DR. NEIL DEGRASSE TYSON, ASTRO-FARMER-CIST

CHAPTER TWO

What Do I Eat in Space?

Human beings evolved on Earth over millions of years, from the rise of our first hominid ancestors to the present day. So did all of our food—and all of the living things that eat our food, and are thus eaten by us by accident. Indeed, every astronaut is a spaceship—home to trillions of microscopic life-forms, all eating and drinking and reproducing. Our digestive systems count on working in a terrestrial environment, so what happens when we bring them with us into outer space?

There are a lot of factors we'll have to take into account, big and small. What do you think, for example, happens when you drink carbonated soda in space, with no gravity in your stomach to separate the gas from the liquid? (Here's a hint: Burps in space aren't very dry.)

So, space gourmands, put aside that pressure cooker and put on your pressure suit. From appetizer to dessert, we're about to find out the best menu for our out-of-this-world journeys.

Feeding astronauts in space has required a lot of culinary creativity—but no food art yet.

Cosmic Cuisine

Why Can't I Get a Pulled Pork Sandwich in Space?

Astronauts can request their favorite menu items, but NASA has to test them. Capt. Sonny Carter, for instance, wanted pulled pork from Georgia, but it failed the test. Of course, Neil was concerned: "I thought nothing survives barbecuing . . . Did you alert him to how many microorganisms he was eating in his pulled pork?" As NASA food scientist Charles Bourland remembers: "Well, he didn't seem to be concerned about them. He'd been eating it all of his life."

It's not always the smartest person who makes a good astronaut. It's personality. You need to be able to live with someone for a long time—be unflappable and easy to get along with because you're confined in a small space for years.

Space food has to be squeaky-clean. Even if they're fully cooked, many well-known foods, especially meats, contain living microorganisms that are harmless on Earth but unwelcome—even dangerous—in spacecraft environments.

But meat has made it into space anyway. In 1965, astronaut John Young snuck a corned beef sandwich onto Gemini 3. By 1989, freeze-dried barbecued pork was on the space shuttle menu.

The best space food is wet—it needs to stick to the plate and fork. "If it's wet, surface tension will keep it on the utensil, so you can eat most things [in space . . . But] it has to be wet," explains Dr. Bourland. "If you open a packet of peanuts, it's all going to float away." ∎

Barbecue fans may find it hard to settle for freeze-dried meat in space.

MARY ROACH:
I tasted seven-year-life-span hash browns.

NEIL:
How were they?

MARY ROACH:
Eh.

THINK ON THIS ▶ Did You Remember to Bring Hot Sauce?

In dry environments with low atmospheric pressure—like space stations—human taste buds and noses don't work that well, so food tastes and smells bland. Chef and TV host Anthony Bourdain knows this well: "Apparently, if you have a stash of hot sauce, you're the go-to guy in outer space. They crave spice." Just be sure your spicy stuff is bacteria-free. A lot of hot sauces are fermented and thus full of microbes—generally a no-no in space.

DID YOU KNOW

South Korean food scientists spent years of research and millions of dollars to develop space-friendly kimchi—spicy fermented cabbage—for astronaut Ko San to bring aboard the ISS.

Cosmic Queries: Space Tourism

Who Wants a Ride on the Vomit Comet?

Enjoy your food in space, but good luck keeping it down. Just ask Neil: "I went into a centrifuge and I barfed up my lunch." What does astronaut Dr. Mike Massimino, expert at keeping his meals down, think of that? "I've never thrown up in a centrifuge. That's pretty sissy . . . I'm talking about in a spaceship."

Zero gravity confuses our digestive systems. Just think about how you feel during those brief near-free-fall moments on a roller coaster—and those puddles often found near the exit gate.

To train for space, NASA previously used a modified turbojet, the KC-135. It created a zero-g environment for 20 to 30 seconds at a time, and a typical mission hit zero g 30 or 40 times. Between 1995 and 2004, NASA cleaned at least 285 gallons of throw-up out of KC-135. The crew nicknamed the jet the Vomit Comet. ■

TOUR GUIDE

Neil's Tips on How to Avoid Motion Sickness in Space

"We live in one g. If that one g changes to any other kind of g, your body reacts. Your ear canals respond, your brain has to try to reconcile what's going on, and in the effort to do so, you get an upset stomach. Motion sickness, it's called. However, once you're in zero g, you can get accustomed to that, because the g-force isn't changing. The first sign of motion sickness is sleepiness. So if the motion sickness is mild, you'll just go to sleep. That means that in space, if you're going from one location to another, see if you can park yourself at that first symptom and just sleep through it."

NASA's C-9 aircraft ascends at a sharp angle.

"With the Vomit Comet . . . they make sure you don't eat for 6, 8, 12 hours before, so now your stomach has got nothing in it to give. So they've reduced—sort of—the vomiting factor, and you just have fun."

—DR. NEIL DEGRASSE TYSON, ASTRO-PUKE-ICIST

Water in space isn't the same as water at home on Earth.

Cosmic Cuisine

What Does Fuel-Cell Water Taste Like?

The water supply on the International Space Station is continuously cycled and recycled—through machinery, out of the air, from people, you name it. "They even recycle any urine from laboratory animals that have been brought up," says astro-pee-icist Dr. Neil deGrasse Tyson. "In fact, the water on the space station is the purest water you would have ever consumed, even if they did extract it from laboratory rat pee."

More than 90 percent of the used dirty water on board the International Space Station is cleansed and turned into superclean water for drinking, bathing, and other uses. One big source of water is from the fuel cells that help power the electrical devices on the station; they produce electricity by combining hydrogen and oxygen gas, leaving water vapor as a waste product.

How does it taste? Might be bland, but . . . pretty much like water. ■

Can You Make a Soufflé in Space?

Gourmet cooking would be tough in a zero-gravity kitchen. What would hold the pan to the stove top, or the baking dish to the oven rack? What would keep the water in the spaghetti pot? What would happen to the droplets of hot oil splattering out of the sauté pan? Things wouldn't be all bad, though. Some foods benefit from the absence of gravity, especially those things that you want to expand during cooking. Imagine, for example, the beautiful stiff peaks you could create by whipping cream; the puffiness of puff pastries; the lightness of meringues; and the fluffy success of that oft-collapsing bane of all home cooks: the soufflé. What else changes in space? ■

DRINK OF THE EVENING

The Martian Sunrise

Concocted by
Neil deGrasse Tyson and
the bartender at the Bell House

1 ½ oz. rum
4 oz. cranberry juice
1 oz. orange juice
Lemon slice for garnish

Fill a highball glass with ice,
then combine ingredients
in the glass. Prop the lemon slice
on the rim of the glass
to represent the Sun.

"If you start shaking a salt-and-pepper shaker, it all goes into the air around you . . . So you'd have to turn all your spices into liquids."

"If you cook a soufflé in space, there is no collapsing under its own weight, because . . . it's weightless."

"I don't know if you want to cook ribs in space, because if you do it right, you would smoke it for 36 hours, and where does the smoke go?"

"Your seasoning would have to be liquid, so that it can stick. Your food has to like itself."

—DR. NEIL DEGRASSE TYSON

THINK ON THIS ▶ What Would Martha Stewart Serve in Space?

For his 10-day space-tourist trip to the ISS, billionaire Charles Simonyi had a meal that was created in part by his girlfriend, domestic doyenne Martha Stewart. It was a freeze-dried feast of quail roasted in mandarin wine; duck confit with capers; chicken Parmentier; apple fondant; rice pudding; and semolina cake with dried apricots.

What's for Dinner?

Astronaut food has evolved from necessary nutrients to real cuisine. You may not recognize their names, though. According to astronaut Dr. Mike Massimino, NASA doesn't like to advertise brands: "We have a certain candy that's very popular because they're small and you can float them, and we call them 'candy-coated chocolates.' "

1962

◄ Astronaut John Glenn, the first American to eat in space, had apple-sauce, pureed beef, and vegetables in aluminum tubes on Friendship 7.

1969

◄ This small stainless steel spoon was used by command module pilot Michael Collins during the Apollo 11 mission. It was part of his personal preference kit.

1973

► The Skylab program introduced refrigerated food and a galley kitchen to space. Astronauts finally got to have ice cream!

1992

▲ A popular choice for astronauts was M&M's, officially known as "candy-coated chocolates" on spacecraft.

2007

◄ TV chef Martha Stewart designed a gourmet freeze-dried meal for her boyfriend traveling to the ISS as a space tourist. The meals were prepared by French chef Alain Ducasse's ADF consulting center.

1965

▶ Project Gemini introduced the first freeze-dried food—a beef sandwich, strawberry cereal cubes, peaches, and beef gravy. At the same time, an astronaut smuggled a deli sandwich into space, but he couldn't eat it because it started to disintegrate.

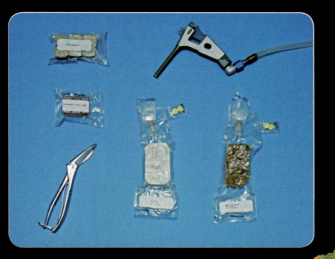

1968

▼ Astronauts celebrated Christmas Day in space with thermostabilized turkey with gravy and cranberry sauce.

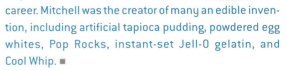

1975

◀ On a joint mission, Russian astronauts shared tubes of borscht (beet soup).

2015

◀ Astronauts made history on the ISS in 2015 when they grew the first successful food in space— romaine lettuce!

FOOD FOR THOUGHT

Did NASA Invent Tang?

Short answer: no. "Tang was on the grocery shelf before NASA was even formed," says Dr. Charles Bourland, a NASA food scientist. Tang was created in 1957 and hit stores in 1959. It first went into space in 1962 with John Glenn—and a culinary myth was born.

The famous drink mix—just add water!—was assembled by the food chemist Bill Mitchell, who earned dozens of patents in his illustrious career. Mitchell was the creator of many an edible invention, including artificial tapioca pudding, powdered egg whites, Pop Rocks, instant-set Jell-O gelatin, and Cool Whip. ■

Astronaut Terry W. Virts tweeted an image of a breakfast burrito he prepared in space.

Cosmic Queries: Human Endurance in Space

Can I Get That on a Tortilla?

All ISS astronauts have a small amount of payload space they can fill with foods of their choice to eat while in orbit. Within the limits of space-approved munchies, the variety is vast—like, for example, astronaut Chris Hadfield's peanut-butter-and-honey tortilla wraps! "All astronauts take comfort food with them," Neil says. "There's meat loaf, there's rice and beans, there's tortillas. Tortillas are great because they don't make crumbs when you eat them." ∎

▶ **HOW QUICKLY WOULD A PIZZA BAKE ON VENUS?**

"A 16-inch pizza would cook in nine seconds on the windowsill." —Neil

9 secs

▶ **HOW QUICKLY WOULD COFFEE FREEZE IN SPACE?**

With an airtight lid, a radiatively cooling cup o' joe would take several hours to freeze solid.

2+ hours

Mouse Stew for Dinner Again?

"It really gets confusing . . . when you need to take any kind of medicine, because it doesn't have the brand names."
—DR. MIKE MASSIMINO, ASTRONAUT

On Mars, agriculture is going to be different. If you choose to raise animals for food, you'll have to get used to different meats than what you're used to finding here in supermarkets on Earth. Sheep, pigs, cows—the typical four-legged farm animals—are large and messy and tough to care for, let alone transport to another planet. Chickens and ducks may be smaller, but their feathers make them even messier. Seafood? Well, without reliable surface water, aquaculture and commercial fishing on Mars don't seem likely either. What's left? "There was this wonderful paper from this 1964 conference on 'Space Nutrition and Related Waste Issues,'" Mary Roach, author of *Packing for Mars*, recalls. "If you were to bring livestock to Mars—like if you're going to bring animals and have ranching going on—what would be the best species to bring in terms of how much it cost to launch them versus how many calories you get? . . . The winner was mice. Mouse stew." So there you have it. The best meal you'll get on the red planet—mouse stew. ■

"You want enough of a diversity, but the bottom line is, you don't need more variation of food than most people would ever have in their own lives. I bet you there's no more than two or three kinds of breakfast cereals you ever eat."
—DR. NEIL DEGRASSE TYSON, TALKING ABOUT ASTRONAUT MENUS

Mars's dry landscape is not suited for livestock.

DID YOU KNOW

A typical house mouse weighs about two-thirds of an ounce, which is about half the weight of an uncooked McDonald's hamburger patty.

LAUGH OUT LOUD ▶ **With Chuck Nice, Comedian**

Here's a question: Can we bring our cats to Mars? Maybe, says Neil: "Pets bring a certain level of tranquility to many people. They form an important psychological support for them in ways that other humans don't. So perhaps." But Chuck Nice has only survival in mind: "So there's your answer . . . You can bring your cat . . . but eventually you're going to have to eat it."

Will Cows Help Us Colonize Mars?

According to Dr. Neil deGrasse Tyson, astro-beef-icist, "A cow is a machine to turn leaves into steak." What does that mean for the red planet? A lot, at least if you ask comedian Eugene Mirman: "Mars is just one cow away from being a place people could live."

But steak isn't the only reason we might want to send cattle to the red planet. Unlike Earth, Mars doesn't have enough greenhouse gas in its atmosphere to keep its surface steadily warm. And cows are famous for producing methane—a potent greenhouse gas. "You'd have to have 10 [cows] for every person, though, because *that* makes sense," explains Dr. Mayim Bialik, neuroscientist, actress, and vegan.

Could cattle help terraform Mars? Actor Paul Rudd certainly thinks so: "I read this article . . . in a magazine—*Life,* I want to say, about 20 years ago—about the terraformation of Mars, and they talked about churning oxygen out of the rocks . . . We would have these kinds of Habitrails, greenhouses, where we would grow food . . . Eventually, you would have cows."

Granted, the current Martian environment would be drastically altered. "Don't we need to damage the environment of Mars to make Mars habitable?" asks actor Michael Ian Black. "Isn't that the entire point?" ■

"I don't think [we'd have farm cows on Mars]. They would need space suits . . . Try milking a cow in a space suit."

—DR. BUZZ ALDRIN, ASTRONAUT

THINK ON THIS ▶ Is There Enough Water on Mars for Agriculture?

Definitely! In one single underground deposit alone scientists have found a huge slab of frozen water—ice that's six times the area of New England and more than 100 feet deep. The challenge would be to melt the water, purify it, and then get it to where you need it to be—all of which would take a tremendous amount of energy and effort.

Packing for Mars (Part 1)

How 'bout I Just Have Some Lasagna?

Of course, astronauts are tough enough that they'd be willing to eat anything they need to survive, but the point of space food is to make the space travelers happy as well as healthy. You would think ice cream would do it, but as food scientist Dr. Charles Bourland explains: "An interesting story about astronaut ice cream is that it was only used on Apollo 8 . . . and they denied eating any of it . . . I don't know if they were too chicken, or if they didn't like it, but later on when we tested it with other astronauts, they didn't like it because it sticks to your teeth."

The kitchen on Skylab, NASA's space station of the 1970s, actually had a refrigerator. There isn't one on the ISS, but the astronauts there still have more than 200 menu items to choose from. "The food actually is pretty good in space, and it's easy to cook. You just add water, you put it in the oven," says astronaut Dr. Mike Massimino. "My favorite was the lasagna. Not quite [like Mama made it] but it was easy to cook [and] it was delicious. We had lasagna, ravioli . . . that's my comfort food. On Sundays I'd have my lasagna, and every other day of the week, [too]." ∎

NEIL TWEETS

Cosmic Eats

"Here's what you do, next Mars mission, bring all the food that has space names to it. No food named Uranus . . . Here we go:"

Eclipse mints

Milky Way bar

Moon Pie

Sunkist

THINK ON THIS ▶ How Do You Make Astronaut Ice Cream?

"Freeze-drying. If I remember my food science, you blow air across the food while it's frozen, and then you evaporate or sublimate the frozen water, leaving behind the flavor and everything else that is the ice cream."
—Dr. Neil deGrasse Tyson, astro-cream-icist

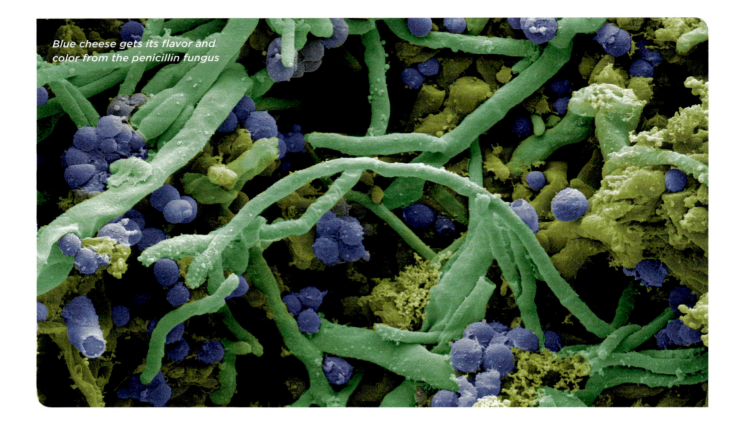

Blue cheese gets its flavor and color from the penicillin fungus

Do You Want Microbes With That?

Pickled vegetables can be good for our gut health.

Decomposition is a totally natural thing on Earth, even when it happens in your refrigerator. Neil explains why: "Why does food decompose? Because there are microbes in it, eating it before you have a chance to get there." And germophobia aside, a little rotting isn't necessarily bad. Indeed, some of the most expensive steaks are dry-aged at room temperature for up to three weeks as they crust over with mold. Before cooking, the surface is cut away, revealing the extra-tender, extra-flavorful pink stuff underneath.

And then there are pickles, sauerkraut, miso, cheese . . . And nutritionists say fermented foods are good for us. ∎

"I visited the NASA 'cosmic kitchen' at Johnson Space Center in Houston . . . I had a steak that had been in a packet on the shelf for five years, unrefrigerated. Some foods they just have to irradiate."

—DR. NEIL DEGRASSE TYSON, ASTRO-NAUT-FOOD-IST

Packing for Mars (Part 1)

Houston, We've Got a Food Problem

I n the early days of spaceflight, NASA astronauts were fed dried cubes of solid food or squeeze tubes of liquid food. The food scientists behind space cuisine at the time weren't concerned with the flavor of the food. Their focus was on nutrition—making sure astronauts got the vitamins, protein, and minerals needed on their missions. As a result, the menu wasn't tasty, and the food delivery systems didn't do a very good job. (Imagine pureeing a hamburger into the consistency of baby food and squeezing it into your mouth like toothpaste.) Astronauts, not surprisingly, complained bitterly—including astronaut Jim Lovell, who directly insulted the meal-creating ability of NASA food scientists while on board Gemini 7. He sent a scathing review of NASA's chicken à la king (see above) back home via NASA's official memo system.

Things got better toward the end of the Gemini era. The squeeze tubes were replaced by plastic pouches of freeze-dried food that could be easily rehydrated. Bite-size cubes came in a plastic container that made reconstituting easier. The new packaging made menu options better, too. Later Gemini astronauts enjoyed choices such as shrimp cocktail, chicken and vegetables, and butterscotch pudding.

By the Apollo era, spoons and forks had become the norm, as did hot water to make rehydrating foods easier. Skylab brought the next big boost in space food. With room for a dining room table, astronauts could choose from an extensive menu featuring 72 different food items. ■

"[On] Gemini 7, Jim Lovell was taking it out on the nutritionist, the guy who invented the food [in the mission transcripts]. 'Memo to Dr. Chance: Chicken à la King Serial Number 654. Cannot even squeeze food through neck.'"

—MARY ROACH, AUTHOR OF *PACKING FOR MARS*

Grilled pork chops aren't on the menu.

DID YOU KNOW
Food for the Mercury and Gemini missions was designed in part to reduce the need for the astronauts to use the bathroom—after all, there *weren't* any bathrooms on those spacecraft.

THINK ON THIS ▶ **Why Not Just Send Plump People to Space?**

"The most extreme approach that I saw was this guy who suggested, and I think he was being serious, that NASA should recruit obese astronauts. He figured out that for 50 pounds of excess weight there's like 184,000 calories. You would just basically starve the astronaut for the duration of the mission." —Mary Roach, author of *Packing for Mars*

"If you take the three-dimensionality of our universe and then you embed it among other universes in another kind of space, that's a higher dimension . . . You step out of your dimension, and then the walls no longer contain you."

—DR. NEIL DEGRASSE TYSON, ASTROPHYSICIST

CHAPTER THREE

Can We Use Wormholes to Travel?

Humans can get to the Moon and back in about three days. To Mars and back, we'd take about three years. To the nearest star system other than ours, though—Alpha Centauri, about 4.4 light-years away—it would take more than 300 *centuries* just to get there. Obviously, we don't have that much time. So what in the universe might we use to cut our travel time?

When we ask that kind of question, what we're really asking is this: What *don't* we know about the universe today that will lead us to tomorrow's physics, paradigms, and, ultimately, technology? Might it be so-called dark matter and dark energy, which together make up 95 percent of the universe's contents? Or might it be black holes?

Let's start with the idea of wormholes—shortcut tunnels through warped space and time. The idea of wormholes has been around for a long time, but always as science fiction. However, confirmation within the past half century of black holes as real cosmic objects has led to brainwork connecting fiction with reality.

Will we one day be able to cross dimensions via a wormhole?

Can You Really Explain the Physics of the Multiverse With an Ant and a Sheet of Paper?

U m, no. Well, maybe—if you're really, really good at origami.

Conceptually, though, yes. Let's imagine a tiny little ant crawling around on a huge sheet of paper. Suddenly, she (it's a worker ant) encounters a wall, stretching left and right without limit. She can't go forward, and she is stuck behind the wall . . . Or is she?

The answer is: Climb up the wall! The ant has left the two-dimensional universe of the paper, which only has left-right or forward-backward directions, and entered a third dimension of space, the up-down direction. At the top of the wall, the ant looks around—and sees her old universe as never before.

Now imagine us humans moving in space. If we reach a wall that extends both up and down and left to right without limit, how do we get past it? If we could climb in a *fourth* spatial dimension, we would no longer be stuck—and then we could see our universe from the point of view of the multiverse! ■

"You step out of our dimension and you get to see things you would not otherwise have even known were there—perhaps even the multiverse itself!"

—DR. NEIL DEGRASSE TYSON, ASTRO-PROFOUND-ICIST

Spaghetti? Me? (Or Neil's Death by Black Hole)

If you're going to visit space, it's best to avoid black holes at all costs. "But what if something awesome is on the other side?" asks Kristen Schaal, actress and comedian.

OK, Kristen, get ready. Here's what will happen when you get too close to a black hole:

▶ STEP ONE

As you fall toward the black hole, the gravity that you experience begins to grow. The gravity at your feet starts becoming greater than the gravity at your head by ever increasing margins. In other words, you're going to get taller.

▶ STEP TWO

The force will eventually exceed the intermolecular bonds that keep your flesh together in one human form. Your body bifurcates, snapping into two pieces, probably separating at your lower spine.

▶ STEP THREE

Meanwhile this tidal force continues, and the two halves of you will continue to stretch. And so your upper half will split into two, as will your lower half. You are now 4 pieces in a train going down, which continues to split, 4, 8, 16, 32 . . . so at one point you will be 210 pieces as you continue to bifurcate.

Comedian Kristen Schaal

▶ STEP FOUR

As you come down to the point that is the center of the black hole, your body funnels down into a narrower and narrower zone. Your pieces will come together left to right, but not top to bottom. So you have been extruded through the fabric of space like toothpaste through a tube. ■

DID YOU KNOW
Supermassive black holes would tear you apart less violently than small black holes as you fall into them. The tidal forces—the difference in the strength of gravity between two points—are not as extreme.

The Science of *Interstellar* With Christopher Nolan

Is a Wormhole Bigger on the Inside?

"The question is, 'Is there any form of energy in the universe that's capable of keeping the wormhole sort of afloat?'"
—DR. JANNA LEVIN, COSMOLOGIST

A wormhole is essentially some kind of warp, channel, or bubble that cuts in and through space-time. Wormholes are pervasive in science fiction, because they represent a way to travel vast distances instantaneously—if they can be controlled.

Fact is, scientists today really think there might be such things. The mathematics of wormholes is highly speculative, but we're learning more about them all the time. "Wormholes, although they are probably theoretically possible, are physically, as far as we know, still impossible," says astrophysicist and author Dr. Janna Levin. "[A wormhole] can be a lot bigger on the inside than it is on the outside . . . To keep the throat open of the wormhole, you need forms of matter and energy that we've never seen before. We don't know anything that could actually keep the throat open on a wormhole. They will keep closing up. It's very unstable."

So the biggest limitation is not the mathematical possibilities, but the energy requirements to control a wormhole. Ten trillion trillion trillion watts of power—the combined output of every star in our galaxy put together—might do it. We're not sure, and it's hard to figure it out in a science lab. ∎

A conceptual computer rendering of a wormhole through space

DID YOU KNOW
The TARDIS of *Doctor Who* fame isn't just a machine that harnesses wormhole technology—it has sentience and intelligence, too.

THINK ON THIS ▶ Could We Control Wormholes Like in the Movies?
"One of my favorite scenes in *Monsters, Inc.*, undiscussed in the reviews of the movie, is that . . . that movie was all about wormholes. Those doors were wormholes . . . wormholes connecting the factory to everybody's closet." —Dr. Neil deGrasse Tyson, astro-worm-icist

Who Discovered Dark Matter?

Even though it's still a mystery, we have known about dark matter for almost a century. "Dark matter was discovered in the 1930s by a dude named Fritz Zwicky," Neil explains. "Back then, it was called the 'missing mass problem,' and it's the longest unsolved problem in modern astrophysics."

Fritz Zwicky (1898–1974) himself couldn't solve the problem. What Zwicky did figure out was that in an area of space called the Coma cluster, galaxies were moving so fast that the cluster should have dispersed long ago. He hypothesized that there must have been more mass in the Coma cluster than what was visible.

At first Zwicky's idea was too outlandish for other astronomers to take seriously. (It didn't help that Fritz was pretty outlandish himself—a bit antisocial, shall we say.) Over the decades, though, the missing mass problem became apparent in more and more places in the cosmos, and today it's widely recognized that *something* has to be there.

"We know that there's dark matter everywhere . . . so we're no longer as shocked when we see things being drawn to one part of the universe or another, just because we don't see anything there."

—DR. NEIL DEGRASSE TYSON, ASTRO-ATTRACTOR-CIST

Things once considered weird—like the Great Attractor, a single area in the universe toward which huge amounts of matter, including our own galaxy, seem to be falling—are now often chalked up to dark matter doing its thing. ■

BIOGRAPHY

👓

Who Is Vera Rubin?

Vera Rubin (b. 1928) pushed the frontier of cosmic understanding from the very start of her career. Two decades before cosmic large-scale structure was confirmed, Rubin concluded in her 1954 doctoral dissertation that galaxies were clustered unevenly throughout the universe. Her work on the rotational motion of the outskirts of spiral galaxies proved that galaxies are embedded within huge halos of dark matter that far outweigh their stars. She was the first woman to make her own observations as an official guest investigator at Palomar Observatory in California, and the second woman ever to be elected to the National Academy of Sciences.

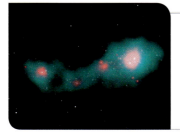

THINK ON THIS ▶ Why Can't We Just Call It Fred?

According to Neil, " 'Dark matter' is really not a good term. We should just call it Fred, because we don't know that it's matter . . . To call it that already puts a bias in the expectation of what you will find, and that's not good science." In fact, the Flash (a DC Comics superhero) has an enemy named Dark Matter, whose "real" name is Fred Fleming. Coincidences and comics aside, though, why Fred and not Fritz? Or Vera?

Cosmic Queries: Answers at the Speed of Light

What Is the Biggest Mystery in Modern Astrophysics?

Let's combine two mysteries into a single biggest one: the fact that 95 percent of the contents of the cosmos is made up of matter and energy that is completely unknown to science.

Observations with telescopes on Earth and in space, at visible light, infrared, and microwave wavelengths, have now painted a picture of the universe with remarkable accuracy—and less than 5 percent of that picture is filled in with things we humans have a handle on, like protons, neutrons, electrons, and neutrinos. Another 25 percent exerts gravity but has no other measurable effect—we call that "dark matter." The remaining 70 percent exerts pressure on space itself to expand and has no other measurable effect either—we call that "dark energy."

Could it be that our understanding of the laws of physics is fundamentally wrong? ■

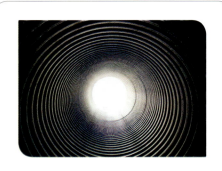

BACK TO BASICS

Are There White Holes, Too?

The idea of white holes—or, more precisely, anti-black holes, where matter and energy come pouring out of a point in space for no reason at all—comes naturally if you think black holes are entry points for matter into cosmic conduits that let out somewhere else. So if wormholes can exist, white holes can exist, too. However, observations show that billions of black holes exist, whereas not a single white hole has been seen. So, rather than being one end of a wormhole, a black hole may be more like a swelling water balloon, with only one way for its contents to come out: the way it got in.

A pulsar shoots a beam of radiation past a planet in its orbit.

"If I were to vote, I would say MOND—MOdified Newtonian Dynamics—is on its way out . . . and we still don't have dark matter completely understood. So until that happens, there'll still be people fighting the good fight—and that's science at its best."

—DR. NEIL DEGRASSE TYSON

Cosmic Queries: New Discoveries

Could a Black Hole Be Destroyed?

Physicist Stephen Hawking first came up with the mathematical formulation showing that, thanks to quantum mechanical processes just inside its event horizon, a black hole can lose mass and shrink over time. How long would it take? A verrrrry loooong time, says Neil: "We call it 'Hawking radiation.' The stuff that goes into a black hole slowly evaporates out of the black hole, until the black hole one day disappears entirely. It's very slow. Takes 10^{100} years to evaporate a super-massive black hole. A googol years."

"Black holes are stronger than any nuclear fusion . . . That's why we got the black hole in the first place. The black hole used to be a star. The star was going to explode, and it didn't. The black hole said, 'No, you're not!'"
—DR. NEIL DEGRASSE TYSON, ASTRO-FUSE-ICIST

Could anything else destroy a black hole? Black holes could crash into one another. In 2015 such an event, more than a billion light-years from Earth, was detected by the burst of gravitational wave radiation it emitted. The black holes weren't destroyed, though—they just merged to make an even bigger black hole. ■

A black hole warps celestial bodies.

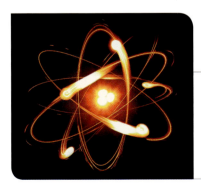

THINK ON THIS ▶ Could Nuclear Fusion Destroy a Black Hole?

Nuclear fusion, the power source of hydrogen bombs, is the most powerful force humans have ever harnessed. Massive stars, the birthplaces of black holes, can generate more energy with nuclear fusion in one-trillionth of a second than the explosive power of all the world's hydrogen bombs put together. And even that's not enough to do *anything* to a black hole.

Exotic Matter in the Universe

What we call "exotic" matter isn't really that weird, cosmically speaking. We just don't encounter these forms of matter under normal circumstances here on Earth, in our narrow environmental ranges of temperature, pressure, and density. And for our well-being, that's probably a good thing! Out in the universe, though, this stuff is all over the place.

||||||||
◀ ELECTRON-DEGENERATE MATTER

In white dwarfs—remnants of stars about the mass of the Sun—gravity presses atoms so close together that a spoonful would weigh several tons. There's some of this material at the center of the Sun right now.

||||||||
▶ NEUTRON-DEGENERATE MATTER

In neutron stars—remnants of stars about 10 times the mass of the Sun—atoms have been destroyed by the intense gravity, and nuclear material is pressed tightly together. A spoonful would weigh several *billion* tons.

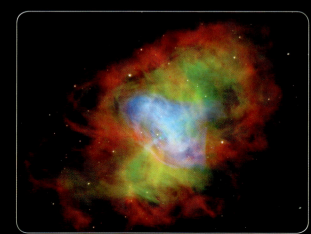

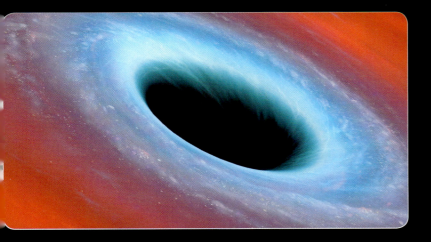

||||||

◀ BLACK HOLES

We have no way to tell what states matter takes within the event horizon of a black hole—nothing can get out from there in a form we can interpret. There's likely to be a singularity at the center that approaches zero volume and infinite density, but it's impossible to be sure.

||||||

▶ STRANGE MATTER

The protons and neutrons in ordinary atoms are made of two types of quarks, called up quarks and down quarks. At superhigh densities, perhaps at the centers of neutron stars, a third kind of quark—"strange"—could hypothetically combine with up and down quarks to create highly unstable versions of the subatomic particles we know.

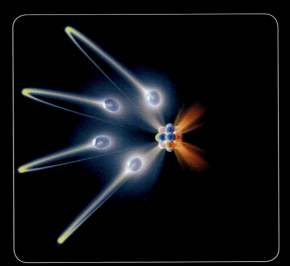

||||||

◀ DARK MATTER

Strangelets, axions, weakly interacting massive particles (WIMPs), huge WIMP-like particles called "WIMPzillas" —none of these hypothetical particles have ever been observed, but they might fit the theoretical requirements of more than 80 percent of the mass in the universe.

What Are the Fundamental Forces of the Universe?

There are four fundamental forces. In order of strength, they are: the strong nuclear force, electromagnetism, the weak nuclear force, and gravity. In quantum theory, each force is transferred by its own kind of subatomic particle—the gluon; the photon; the W^+, W^-, and Z^0 particles; and the graviton.

WEAKLING FORCE

We still don't know why gravity is so weak compared with the other forces. As Neil puts it: "Gravity isn't just the weakest; it's the stupidly weakest force." Think about it: The tiny bit of electromagnetic force from the static electricity in your hair can hold a balloon on your head—defying the gravity produced by the entire Earth.

SAVING GRACES

On the other hand, gravity has two things going for it. Unlike the nuclear forces, gravity can work over long distances; and unlike electromagnetism, there are no positive-gravity and negative-gravity particles that cancel each other out. So on a cosmic scale, at cosmic distances, gravity shapes the universe without peer.

QUASARS

A quasar is an example of a cosmic gravitational engine—a supermassive black hole surrounded by matter falling onto it, located at the centers of galaxies. Quasars can produce more energy in one second than the Sun could in 10 million years! ■

$$F = ma$$

Cosmic Queries With Bill Nye and Astro Mike

How Big Can a Star Get Before It Becomes a Black Hole?

Stars produce black holes when, late in their lives, their inward gravitational pull overwhelms their outward energy output. In stars, mass and size don't always correspond; the Sun, for example, will never make a black hole, but for part of its 10-billion-year lifetime it'll be bigger than a more massive star that would make one.

"So would you describe it as over a million whales?" asks comedian Eugene Mirman. Sure, we can use whales to try to get a handle on stars. The sun has a mass of 10 million million million million adult blue whales! That's big, but not big enough. If a star is less than 8 times the mass of the sun, it's probably not going to make a black hole. If it's more than 20 times, it'll probably make one. In between, it could go either way. ∎

BIOGRAPHY

👓

Hubble the Racist or Hubble the Constant?

Edwin Hubble (1889–1953) transformed humanity's understanding of the universe. Tall and handsome, he played sports at the University of Chicago, earned a Rhodes scholarship to study law at Oxford, and ultimately followed his love of astronomy to Southern California. He was an ardent social climber, putting on British upper-class pretensions although he was fully American; in conversation, he and his wife often used racial slurs common of the time. At the observatory, though, there was no better astronomer of his era. He proved the existence of galaxies far beyond the Milky Way, and showed that the universe is expanding—the result of the big bang.

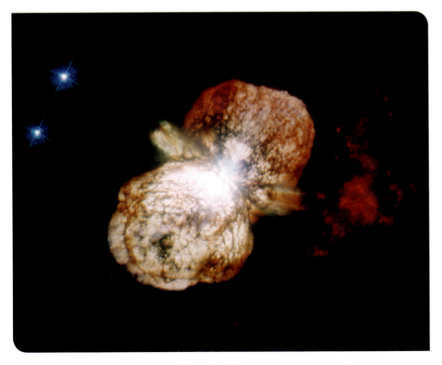

Hubble captures gas and dust billowing from a supermassive star.

"We see black holes in the center of every galaxy we've had the power to look into . . . and when you do, you can say, 'I bet you were a quasar back in your youth.'"

—DR. NEIL DEGRASSE TYSON

Let There Be Light

What Does Cosmic Background Radiation Reveal About the Early Universe?

Light travels at a finite speed—so in the same way that a postcard sent from far away is a record of the past, the view of a faraway object is a record of what it looked like when light left that object. That's called look-back time. Astronomers use it to measure the history of the observable universe. Carter Emmart, an astronomer, tells us how this works: "If you look far enough, you're actually seeing the cooling off of the universe, or the transition between when it was a plasma, and opaque, to where it became clear space. And that's the microwave background that we see." ■

The universe has been slowly cooling since the beginning of time.

BACK TO BASICS

How Did the Big Bang Lead to Matter—and Us?

"At the big bang, you have all of this energy in a very small volume, and it is so hot, it is so energetic, that matter is forming out of the energy, according to $\mathcal{E} = mc^2$. So what we have here is this soup; it's a matter/antimatter energy soup. And then as it expanded and cooled, all the matter and anti-matter particles collided with each other and annihilated and created light. Except some matter was left over out of this—one-billionth of the particles was left over—and that is the matter that you and I are made of. And all the other matter and anti-matter collisions created the light of the universe, now visible as microwaves at the most distant regions of the universe." —Dr. Neil deGrasse Tyson

"If you go away from Earth far enough, you're going far enough back in time. Everything you see—a star twice the distance away is twice as far in the past, essentially."

—CARTER EMMART,
ASTRONOMER AND ARTIST

*"The problem with that is . . .
we have no clue how to go that fast."*

—DR. PHIL PLAIT, ASTROPHYSICIST
AND AUTHOR OF *BAD ASTRONOMY*

`Cosmic Queries: Time-Keeping`

Are Photons Timeless?

You could say, in a manner of speaking, that photons are timeless. Leave it to Neil to explain: "As you travel faster and faster and get closer and closer to the speed of light, you age more and more and more slowly. Time ticks more and more slowly for you. If you hit the speed of light—which we don't know how to do yet, but . . . if you ever hit the speed of light, then time stops. Photons, which are the carriers of light, exist at the speed of light. A photon doesn't accelerate from zero to the speed of light in 2.4 seconds; it exists at the speed of light. Because it exists at the speed of light, any watch that it is carrying never ticks . . . If you are a photon and you are emitted from across the universe, you will be absorbed, you will slam into whatever you were destined to hit, as far as you are concerned, instantaneously. No time would have elapsed." ■

TOUR GUIDE

What Happened While You Were Off Exploring the Universe?

One fascinating consequence of Albert Einstein's theory of relativity is time dilation. Whereas we may feel like time passes at the same rate for everyone and everything, it's actually different for moving objects—and very different for objects moving very fast. "If you were to get in a spaceship and move very close to the speed of light, you could literally travel across the galaxy, and somebody on Earth would experience a hundred thousand years elapsing while you might only experience a few months elapsing," says Dr. Phil Plait, astrophysicist and author of *Bad Astronomy: Misconceptions and Misuses Revealed*. "And that is time travel into the future—because if you went to the star, poked around, planted your flag, turned around, came back, 200,000 years would have elapsed on Earth, and you would have only experienced a fraction of that."

THINK ON THIS ▶ Why Does Time Slow Down If You're Moving Fast?

"You really have to say, 'No, it's not just that you think time is slowing down.' Your watch is physically ticking at a different rate than a different watch . . . Subatomic particles are decaying and doing whatever it is they do on a daily basis behind closed doors, [but] they're doing it at a different rate."
—Dr. Phil Plait, astrophysicist and author of *Bad Astronomy: Misconceptions and Misuses Revealed*

"We explore the solar system and the rest of the cosmos with our robots, which are basically our eyes and our ears. So it's great: I get to go explore the cosmos from the comfort of my couch, which I love. I can still eat doughnuts . . . It's a much better life."

—DR. AMY MAINZER, ASTROPHYSICIST

CHAPTER FOUR

Who Goes There?

As humanity takes its first steps beyond its home planet, what are we sending (who), how are they traveling (goes), and where are they going (there)?

Right now, we're sending robotic spacecraft, filled with scientific sensors and communications equipment. What has their use taught us about space—and about ourselves? They're extensions of us, after all, our remote eyes and ears. If these robots also had brains, would they still be "us"—or would they become "them"?

Robots or humans, our intrepid explorers are looking for wonders yet unimagined. Maybe the most commonly imagined wonder, though, is extraterrestrial life! We think we're most likely to find simple life-forms—unicellular microbes, or algae, or primitive plants and animals. The chance is there, though, that we'll find intelligent creatures—who, upon seeing us snooping around the cosmos, might well indeed ask us: Who goes there?

NASA's robot Valkyrie is six feet tall and weighs 290 pounds.

Do Scientists Really Look Like Their Spacecraft?

Every space probe embodies the work of hundreds or even thousands
of people and their collective blood, sweat, toil, and tears.
So, like Dr. Frankenstein in Mary Shelley's classic novel,
are these scientists and their visages revealed
in the form of their creations?

‖‖‖‖

◀ WIDE-FIELD INFRARED SURVEY EXPLORER (WISE), DR. AMY MAINZER

Launched in December 2009, WISE is a four-channel supercooled infrared telescope. The telescope's mission is to survey the entire sky, with 1,000 times more sensitivity than all previous infrared missions.

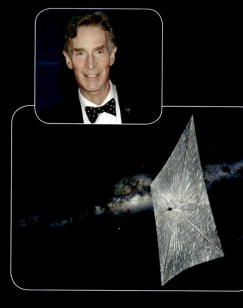

‖‖‖‖

▶ LIGHTSAIL, BILL NYE

Built with reflective sails, LightSail completed a test flight in June 2015 and is expected to set sail again in 2017. The spacecraft—a citizen-funded project by the Planetary Society—uses the Sun's energy for propulsion—flight by light.

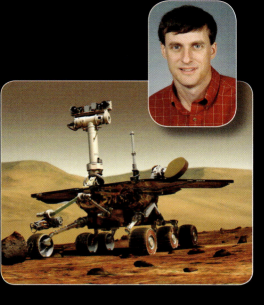

◀ SPIRIT AND OPPORTUNITY, DR. STEVE SQUYRES

As part of the Mars exploration rover mission, Spirit traveled 4.8 miles on the surface of Mars between 2004 and 2010. Opportunity, its twin, has traveled more than 25 miles since 2004.

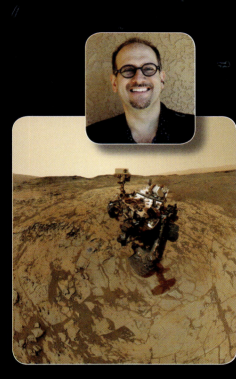

||||||

◀ CASSINI, DR. CAROLYN PORCO

Launched in 1997, Cassini entered Saturn's orbit in 2004. Its mission includes studying the planet's rings, atmosphere, and moons.

||||||

▲ CURIOSITY, DR. DAVID GRINSPOON

Curiosity landed near Mars's Gale crater on August 6, 2012. The rover was still operational as of spring 2016.

"So, literally, aliens can come to Earth, say hello in 55 languages, and then murder people. Literally, 55 nations will let their guard down, maybe more."

—EUGENE MIRMAN, COMEDIAN

`StarTalk Live!: I, Robot (Part 1)`

What's So Special About Voyager 1?

n 2012 the Voyager 1 space probe entered interstellar space—the first human creation to do so. Its instruments recorded a change in the environment around it, proving it had crossed the threshold of the sun's electromagnetic influence. It now takes light and radio waves nearly 20 hours to get from Voyager 1 to Earth, and the spacecraft is now more than 12 billion miles from Earth.

"What's special about Voyager is that it was launched with enough energy to careen around Jupiter and Saturn," Neil says. "By the time it exited the solar system it had enough speed to leave the solar system entirely, and upon doing so, just recently it actually crossed the border between our solar system and space. The farthest object we have ever sent anywhere, ever."

And it's still working great! NASA's plan is to continue using Voyager to obtain science data on interplanetary and interstellar fields, particles, and waves until at least 2020. All this with a machine whose available onboard computer memory is about .000002 gigabyte—one ten-millionth that of a typical smartphone. ■

Voyager 1 launches aboard a Titan rocket in 1977.

Haley Joel Osment and Jude Law play humanoids in A.I. Artificial Intelligence (2001).

Are Robots "Them" or "Us"?

An industrious Wall-E toy from Disney-Pixar

The philosophical definition of what makes us human doesn't include distinctions between biological or mechanical bodies. We may make "them" mechanically, but we make copies of "us" biologically. So what if they're not as advanced as we are yet? Human reproduction took four billion years to evolve, and we've only been making robots for about four decades.

And what do we mean by "advanced," anyway? Without technology, humans aren't advanced enough to travel deep into interplanetary space, or roam the Martian surface without food, or analyze Saturn's magnetic field. But our robots can. ∎

"We shouldn't dismiss these robots as saying that that's not like sending humans, because we are . . . We outsource our cognition to these robots. The mind of man crawls Mars at the moment."

—JASON SILVA, FUTURIST

*"When it manipulates things,
I think that's what makes it a robot."*

—STEPHEN GOREVAN,
ROBOTICIST AND SPACE SCIENTIST

StarTalk Live!: I, Robot (Part 1)

What Is a Robot, Really?

DRINK OF THE EVENING

The Robot Cocktail

Concocted by Neil and the
bartender at the Bell House

A splash of soda water
½ glass of pineapple juice
3-count of Grand Marnier

Pour all of the ingredients,
in order, over ice. Enjoy!

Like so many words, the word "robot" means something in day-to-day conversation that is different from its technical usage. Creations we would certainly call robots today were imagined even in ancient times. The Greek god Hephaestus, it is said, built metal automatons to serve him. According to Greek mythology, by the way, Hephaestus also constructed Pandora, the first woman. In any case, robots—of either the hardware or software variety—need to have a certain level of complexity, flexibility, and in some cases even a capacity to learn. Even those terms, though, have varying definitions, depending on whom you ask.

Here's roboticist and space scientist Stephen Gorevan's definition: "A robot is a programmable manipulator. Now I realize that sounds more like a cross between a computer and my ex-wife, but if a machine doesn't do more than a few things, it's not really a robot . . . If it does sort of one thing or a short classification of things, then it's an automated spacecraft." ■

*"If a person introduced
the robot to me as a human,
I would deal with it that way,
because I don't know
what this person's deal is.
I don't want to judge."*

—JASON SUDEIKIS, COMEDIAN

LAUGH OUT LOUD ▸ **With "Mars Czar" Dr. G. Scott Hubbard**

"It's very difficult to compare the generic robotic mission with the generic human exploration mission, but sending humans, because of the life support and all the other issues, is at least 10 times, maybe as much as 100 times, more expensive . . . Assuming you want to keep them all alive."

Bill Nye joins Neil at the premiere of Cosmos: A Spacetime Odyssey.

Where's the Sense of Adventure?

One day, we all may carry these passports as we travel through space.

We want to live vicariously through our adventure heroes, and imagine that we're making these exciting scientific discoveries and pioneering things ourselves. Is achieving these tasks by machine simply not as good as doing so in person?

Planetary scientist Steven Squyres may think so: "I have spent the last 20 years of my life trying to design and operate robots that can replicate what a human might be able to do on the Martian surface. What our rovers do in a day, you and I could do in about 30 seconds . . . Humans have a capability to synthesize information, to digest it, to figure out the next thing to do, and to improvise. Robots can't improvise the way humans can."

So, yes, sooner or later we really should send people to Mars or wherever we want to study. "Humans [also] have a capability to inspire that robots simply lack," says Squyres. "Someone once famously said, 'No one's ever going to give a robot a ticker-tape parade.'"

But until we can go safely, robots will pave the way. ∎

How Good Are the Russian Robots?

A Soviet Union propaganda poster says its men opened the road to the stars.

NASA brought back more than 800 pounds of moon rocks but had to send people to get them. The Soviet space program, on the other hand, brought back lunar samples completely robotically. And, according to astrobiologist David Grinspoon, also known as Dr. FunkySpoon, there's more: "The Russians couldn't land a spacecraft on Mars to save their life. They couldn't explore their way out of a paper bag on Mars. They threw all these spacecraft at Mars and they failed. It's actually very sad the amount of resources the Russians put into Mars.

"But for Venus, they were really successful. They had orbiters, they had incredible landers—the first landings on Venus were done by the Russians. And all those pictures you've seen—pictures of these strange landscapes, with rocks going off into the horizon of this eerie world, the first pictures we got anywhere from the surface of another planet—were done by the Russians with these really successful and incredibly well-engineered landers on Venus." ■

"Do you think that what America always needs is the threat that someone else might do something first? Because it seems like part of the space program that you were able to do—well done, by the way—was in competition with the Russians."

—JOHN OLIVER, ACTOR AND COMEDIAN

LAUGH OUT LOUD ▶ **With Eugene Mirman and Neil**

A thick layer of ice on the surface of Jupiter's moon Europa prevents us from studying its vast underground oceans. How might we get through?

NEIL: A human would probably not be the first to attempt to get through that ice.

EUGENE MIRMAN: You'd send a bunch of cats that you would teach how to dig.

What Led to the *Challenger* Disaster?

O n January 28, 1986, the space shuttle *Challenger* broke apart during its ascent into space. All seven astronauts aboard the flight were killed in the crash.

A lengthy investigation revealed that people at just about every decision-making level, from the contractors who built the space shuttle and its parts to the NASA officials in charge of the launch, made crucial errors in judgment and communication that led to the disaster. "If you have a society which insists on pointing the finger of blame at individuals every time something goes wrong, you're going to create behavior on the part of the individuals that is risk-averse, that is defensive," says author Malcolm Gladwell.

Sadly, without herculean efforts from leadership and management, cultural issues like this can develop in just about every corporate and bureaucratic culture. ∎

The space shuttle Challenger *exploding shortly after liftoff*

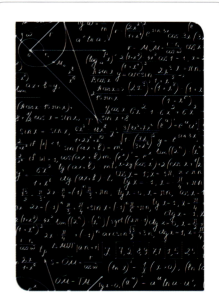

BACK TO BASICS

Why Is Rocket Science So Hard?

Rocket science deals with a lot of energy being released, a lot of force being applied, and a lot of mass being moved a long way. One seemingly small error can have huge consequences. Here are just three examples:

2003: The space shuttle *Columbia* burns up during reentry into Earth's atmosphere, killing all seven astronauts aboard. The leading edge of the left wing was damaged during launch, causing the orbiter's thermal protection system to fail.

2011: The Russian probe Phobos-Grunt, designed to bring rock samples back from a Martian moon, gets stuck in low-Earth orbit and crashes into the Pacific Ocean two months after launch. At least one rocket failed to fire properly.

2015: A SpaceX Falcon 9 rocket disintegrates in midair two minutes after launch. A single support strut failed, causing catastrophic structural failure of the entire rocket.

Cosmic Queries With Bill Nye and Astro Mike

How Do Spacecraft Move in Space With Nothing to Push On?

A vintage tin rocket racer toy

One of the most common misconceptions about space travel is that in order to move, you have to push off something, like Earth's surface. "When you watch the rocket leave the ground, it gives you the impression that the flames and gases are pushing against the Earth, but that's not really what's going on," explains Bill Nye the Science Guy. "You're throwing hot gas out the back of the rocket so fast that the reaction is the rocket goes off in the other direction . . . This works whether you're on the Earth or in space."

Newton's third law of motion—for every action, there is an equal and opposite reaction—means that the simple act of pushing is enough; you don't need to push *on* or *off* any external object or surface to generate motion.

Some rockets and nonrocket systems, like the Hubble, use reaction wheels to move. "You spin them in a certain direction and you get a reaction in the other direction," says astronaut Dr. Mike Massimino. "Then the thing will point where you want it." ■

> "We're going to have this very large sail deployed from this . . . tiny spacecraft. . . . You steer it by tacking, the same way you steer a sailing ship."
> —BILL NYE, THE SCIENCE GUY, SPEAKING OF LIGHTSAIL, A SUNLIGHT-POWERED SPACECRAFT

> "People worried . . . suppose [it] enters our atmosphere and then disintegrates, then it scatters plutonium around the world, killing everyone. So there were some protests at the time . . . It didn't happen, because we know Newton's laws of motion. We got this one."
> —DR. NEIL DEGRASSE TYSON, ASTRO-NUKE-ICIST

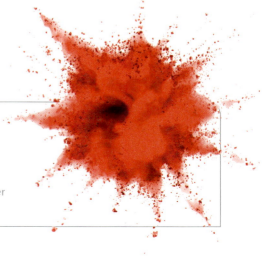

THINK ON THIS ▶ **Have We Been Sending Nukes Into Space?**

Some nuclear reactors have been sent into orbit, but not yet on scientific space probes. Radioactive materials, on the other hand, have been used often; their radioactive decay produces the small amount of steady heat used by radioisotope thermoelectric generators (RTGs) to power the probes' electrical systems. The RTGs are heavily shielded and have never endangered humans on Earth.

Cosmic Queries: Tour of the Solar System

Why Did Lowell Think There Was Life on Mars?

Percival Lowell (1855–1916) was the black sheep of a wealthy Boston family, the brother of poet Amy Lowell and Harvard president Abbott Lowell. He founded Lowell Observatory in northern Arizona—still a world-leading research institution today—and studied Mars from there for two decades.

Lowell is famous for announcing that he observed canals engineered on the surface of Mars. "People looked through a telescope a hundred years ago, and in moments of brief atmospheric clarity they could see what looked like a fine network of straight lines on the planet's surface," says planetary scientist Dr. Steven W. Squyres. "And they were so straight and so regular that the people looking at these concluded that not only were they evidence of life, but they were evidence of intelligent life. Now they were correct, but the life was at the wrong end of the telescope. What they were seeing was an optical illusion. In fact, there's nothing of the sort on the surface of Mars."

Lowell's mistake didn't make him any less of a scientist, though. His evidence was flawed, but his efforts were earnest, and mistakes are part of the process of discovery. Lowell would have been a bad scientist only if he had clung to his incorrect position against clear scientific evidence— which, at the time, was unavailable. ∎

TOUR GUIDE

Will We Find Life Inside or Outside Our Solar System?

The exploding number of known exoplanets—planets outside our solar system, orbiting stars other than the sun—has made some scientists think we'll have better luck looking beyond. There are, after all, thousands of exoplanets and only one Mars! That said, Neil still thinks we'll find life on Mars first: "On Mars, it's the microbial life we're eager to find. . . . If we find life on exoplanets, we're not looking at it microbially . . . We're at the cusp of being able to look at the atmospheric chemistry of planets that are orbiting other stars . . . We'll find life sooner on Mars than on an exoplanet, just because the technology isn't completely there yet."

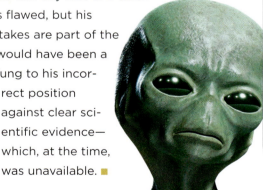

LAUGH OUT LOUD ▶ With Neil, Kristen Schaal, and Eugene Mirman

NEIL: How do you define life? If a beam of light came across and it was life, we probably would not define it as such.

KRISTEN SCHAAL: No, we would just bask in it.

EUGENE MIRMAN: We would bask as it secretly got us pregnant.

Could Life on Earth Have Originated on Mars?

Here's what we know so far according to Neil: "We learned only recently that when an asteroid hits, surrounding rocks can be thrust back into space and recoil with sufficient ferocity that they can escape the planet altogether . . . And, we have found, if you have a rock with nooks and crannies, and if the originating planetary surface is rich with life, and that rock escapes the planetary surface, it can have stowaway microbes that can survive the vacuum of interplanetary space. So the rock moves through space and lands on another planet. We have a word for this—it's called panspermia."

The tastefulness of the term aside, panspermia is only possible if the microbes survive a brutal interplanetary journey—and one that lasts millions, if not billions, of years. Experiments and computer simulations suggest a very, very low probability of such survival. On the other hand, the entire microbe may not have to survive—maybe its molecular constituents alone, like protein or RNA or DNA, would do the trick.

"So here's the thing," says Neil. "Evidence shows that Mars, as a planet, was wet and fertile earlier than Earth was. That being the likelihood, a rock from Mars could have been infused with microbes from there and, landing on Earth, it could have spawned life as we know it. Which would tell us then that all life on Earth descends from Martians." ∎

The fiery crash landing of an asteroid on Earth

What's Biting on Europa?

Jupiter's moon Europa, slightly smaller than Earth's moon, is covered with a thick layer of frozen water. Its surface is scarred with ridges, cracks, and seams, much like Earth's polar ice caps. Deep below this icy surface, could there be a vast underground liquid ocean? Could there be life? We're looking for it, according to astrobiologist Dr. David Grinspoon: "Europa is NASA's top priority for a big mission, because it's one of the places where there ought to be life, if we're right about what it takes for life. There's an ocean, we think, beneath this icy crust. In fact, it may be our solar system's biggest ocean of liquid water."

"I want to be the first person to eat a space lobster. Deadliest Catch: Europa."
—EUGENE MIRMAN, COMEDIAN

It's not a sure thing, of course; life exists in Earth's oceans under thick ice caps, but that water has been oxygenated by eons of photosynthesis by blue-green algae. There's no simple way for that to have happened on Europa. Still, the allure is powerful: We want to find out for sure. How are we going to do it?

"The engineering problems are so titanic, NASA is smart enough to know . . . we're going to need decades of research and development," explains Dr. Stephen Gorevan, co-founder of Honeybee Robotics. "We're talking about digging a hole . . . millions of miles away, with only a small amount of power . . . half a mile deep . . . And there's no solar power available out there, so you're going to have to bring nuclear power." ■

BACK TO BASICS

Does Salty Water Mean There's Life Down There?

Scientists generally agree that to sustain life as we know it, an ecosystem must have a steady heat source, liquid water, and a critical set of chemicals, including carbon and nitrogen compounds. Finding all three in one place, like Saturn's moon Enceladus, makes it a hot spot to look for extraterrestrial life. "[Enceladus] is exactly the kind of environment that . . . could be inhabited by living organisms," says planetary scientist Dr. Carolyn Porco. "It's watery. The salt in it tells us that the water is in contact with rocks, so there's available chemical energy for the organisms to live if they can't live off sunlight. And there's organic materials, so to me, it is the most accessible habitable zone in our solar system."

THINK ON THIS ▶ What Would Life Look Like on Titan?

Saturn's moon Titan is even larger than the planet Mercury, and it also has an atmosphere thicker than Earth's. But it's cold—its mountains are made of frozen water, and its lakes and rivers are made of liquid natural gas. If life could exist there, it could be truly exotic by Earth standards. Maybe, suggests Bill Nye the Gas Guy, it ingests hydrogen and acetylene while excreting methane.

"Welcome back to StarTalk, from Manhattan, New York City, North America, Western Hemisphere, Earth, solar system, Sagittarius Arm, Milky Way galaxy, Local Group, Virgo supercluster, the universe . . . And we don't yet have a coordinate within the multiverse—we're working on that."

—DR. NEIL DEGRASSE TYSON

CHAPTER FIVE

Are We There Yet?

We don't have to go far—just a few hundred million miles, well within our own solar system—to find all kinds of fascinating objects, structures, and materials. Not just big things, either—everything from chunks of ice, rock, and metal, from small planetlike bodies down to building-size blocks.

The vast majority of the mass in the solar system—more than 99.99 percent—resides in just 10 massive bodies. But the number of small solar system objects—at least 10,000 at last count (and that's probably less than one percent of what's out there)—far outstrips the number of big planets and moons. That's plenty of new places to examine and explore, right?

Our first explorations of these small solar-system bodies may bring opportunity! Many of the first voyagers to the "New" World set out to seek fortunes. In the same way, discovery and exploitation of what's out there in the solar system could bring economic wealth. How will that change the "Old" World we live on now?

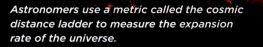
Astronomers use a metric called the cosmic distance ladder to measure the expansion rate of the universe.

Why Is Kepler-37b a Planet, but Pluto Is Not?

Well, rigorously speaking, Kepler-37b is an exoplanet and Pluto is a dwarf planet—so they're both sorta-kinda planets and both sorta-kinda not. What's in a name, right? Here are just a few very interesting planetlike bodies.

▲ PLUTO

Pluto has a radius of 740 miles. In 2006 the International Astronomical Union (IAU) downgraded it to a dwarf planet. While it orbits the sun and is nearly round, Pluto does not meet the third requirement of a planet; it has not "cleared the neighbourhood around its orbit." Or, as Neil puts it: "If Neptune were a Chevy Impala, then Pluto would be a Matchbox car, at best."

▲ MOON

Orbiting approximately 238,000 miles from Earth, the Moon (radius of 1,080 miles) is the only other solar system object visited by humans. In total, 12 astronauts have taken the three-day journey to the moon.

▲ KEPLER-37B

Discovered in 2013, exoplanet Kepler-37b is one of the smallest planets yet discovered around a main-sequence star. Though it has a radius only slightly greater than the Moon's, it is considered a planet because the IAU's definition only applies to our own solar system.

▲ MERCURY

The closest planet to the Sun, Mercury is one-third the size of Earth, with a radius of 1,520 miles. The planet's large metallic core makes up about 80 percent of Mercury's radius. Daytime temperatures on Mercury can be six times hotter than the hottest spot on Earth.

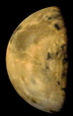

▲ MARS

The red planet has a radius of 2,110 miles—about half the diameter of Earth. Mars is a cold desert planet, but much like Earth it has seasons, volcanoes, and varying weather. However, Martian gravity is only 38 percent of that on Earth, and its atmosphere is too thin to sustain liquid water on its surface.

▲ EARTH

Our home planet has a radius of 3,960 miles and is the only planet in our solar system that we know of to host life. Earth orbits the Sun—at a distance of 93 million miles—and scientists have measured its age to be more than 4.5 billion years.

▲ KEPLER-37D

Discovered with the Kepler Space Telescope, Kepler-37d is the largest of the three known exoplanets orbiting the star Kepler-37. It is twice the diameter of Earth and orbits its star every 40 days.

How Is Jupiter the Solar System's Vacuum Cleaner?

The biggest planet in our solar system, Jupiter takes care of us. Neil explains how: "Jupiter has more mass than everything else known that orbits the Sun. So, if you're a comet coming from far out in the solar system and you're aiming towards Earth, you've got to get past Jupiter. And here's your trajectory, with Earth in your sights, and Jupiter says, 'Uh-uh. First, I want you to hit me.' And Jupiter can eat a comet like it's nobody's business.

"Others will come in, try to get past Jupiter unscathed, and they won't—Jupiter will swing the thing around, do-si-do, and cast it back out to the solar system so it never even makes it to the Sun. And in other cases, it can swing around Jupiter and be flung out of the solar system entirely.

"[Jupiter] is our big brother protecting us from the hazards of the outer solar system . . . If it weren't for Jupiter, you can justifiably question whether Earth could have ever made it from simple life to complex life." ■

Comet Shoemaker-Levy 9 on a collision course with Jupiter

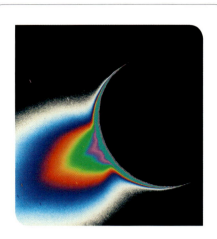

BACK TO BASICS

Where Did Saturn's E Ring Come From?

So far, in the 21st century we've discovered two rings of Saturn well beyond those ordinarily visible with a small telescope from Earth. And one of those rings happens to have one of Saturn's most interesting moons orbiting within it: "That beautiful blue E ring . . . is created by a hundred geysers erupting from the south polar terrain of a tiny moon called Enceladus, which is no bigger across than Great Britain," planetary scientist Dr. Carolyn Porco says. "Those geysers, we are virtually certain, erupt from a layer of salty liquid water laced with organic materials and bathed in excess heat."

LAUGH OUT LOUD ▶ With Dr. Neil deGrasse Tyson, Astro-Tub-Icist

"Saturn's density is so low because it has so much gas in it that . . . if you scooped out an average piece, it would float on water. And when I was a kid, I wanted a rubber Saturn instead of a rubber ducky—because I knew Saturn floated—to play with in my tub. But no one made it."

Where Do Comets Come From?

Astronomers divide comets into two general categories—short-period comets, which take less than a couple of centuries to orbit the Sun, and long-period comets, which take more. Most short-period comets reside in the Kuiper belt—a doughnut-shaped zone beyond the orbit of Neptune, centered on the sun, that also contains Pluto, Eris, and several other dwarf planets that are essentially really big comets.

Most long-period comets reside in the Oort cloud (named after the Dutch astrophysicist Jan Hendrik Oort), a huge spherical shell also centered on the sun. It may be a trillion miles across and may contain many trillions of comets. A very tiny percentage of Oort cloud objects ever make it to the inner solar system.

Taking this question in a more existential direction—that is, how do comets come into existence?—we face a big cosmic mystery. We know comets are made of ice and dust, but how did scads of little bits of matter come together in deep space to make these distant, isolated solid chunks?

"Comets and some other asteroids might be rubble piles, just rocks that are gathered together," Neil tells us. "It's porous, or it's just a pile of rubble traveling together, pretending it's one solid object." ∎

> "It was not until recently that we had any clue what the structural integrity of comets or asteroids actually is . ... We don't really know how tightly held together these things are."
>
> —DR. NEIL DEGRASSE TYSON

Comet McNaught blazed across Australia's lower Eyre Peninsula in 2007.

An asteroid belt around bright star Vega

How Did the Asteroid Belt Get There?

▶ **HOW MUCH MASS IS IN THE ASTEROID BELT?**

"Bring all the asteroid pieces together, and they will sum up to about 5 percent of the mass of our moon." —Dr. Neil deGrasse Tyson, asteroid-icist

5%

"Asteroids and comets are kind of like the leftover bits of junk that were the remnants of what happened when our solar system first formed, about 4.5 billion years ago," explains astrophysicist Dr. Amy Mainzer. "Those little bits of space debris are what became the asteroids and the comets today. Things that formed a little further away from the sun, where it was cold and really dark—those formed the comets, the icy stuff. And the stuff that formed closer to the middle, the rockier stuff that formed where it was too hot for ice—that's how you get asteroids." ■

DID YOU KNOW

Astronomers think much of Earth's oceans could have been formed by comets striking our planet and depositing their water on Earth's surface.

Can a Solar System of Stars Exist?

DID YOU KNOW
"The average speed of most asteroids and comets is about 20 to 30 kilometers per second, or about 40,000 to 60,000 miles per hour," says Dr. Amy Mainzer, astrophysicist. Earth, meanwhile, zips around the Sun at 66,000 miles an hour.

Look up at the night sky and you begin to imagine variations on our home solar system. Does each star have its own system of planets? As Neil puts it: "Half of the stars you see aren't solo stars at all. They're double, multiple, triple, quadruple star systems. Even, for example, the nearest star to the sun, Alpha Centauri, that's a multiple star system."

Technically, a multiple star system isn't a solar system, but planets can orbit multiple star systems. In other words, a planet like Luke Skywalker's Tatooine is possible. Planets can even swing between the stars, orbiting one star for a time, and then another later. "If they are far enough away from one another, then each one can have its own planetary system," Neil explains. "But if they're too close to one another, as the planet comes around the back stretch, its gravitational allegiance can be compromised." ■

The rotation rate [of Comet 67P] is once every 12 Earth hours . . . twice as fast as the Earth spins. It's a difficult deep space problem to get your Rosetta probe rotating at about the same speed.
—BILL NYE THE SCIENCE GUY

Star system HD 98800 consists of two pairs of double stars

"In deep space when you have dust many, many kilometers apart, and they're being attracted, they actually slam together at a pretty high speed."
—BILL NYE THE SCIENCE GUY, ON WHY COMETS ARE HARD TO MAKE

LAUGH OUT LOUD ▶ With Chuck Nice, Comedian
As Neil explained, when stars are near each other, the planets orbiting them might become confused by a star's gravitational pull. "Ah," says Chuck Nice, "so it just doesn't know, 'Which one do I belong to? I want to go with Proxima, but I just can't help myself. Alpha's just so sexy.'"

Asteroid, Dwarf Planet, or Both?

Right now, Ceres is both the largest asteroid and the smallest dwarf planet.
In life, lots of things occupy two or more classifications,
so why not solar-system objects?
That doesn't mean everyone has to agree, though.

|||||||

▶ DWARF PLANET

If a solar-system object is roughly round and also orbits the Sun as its primary, but is not by far the largest object in its orbital path, then the International Astronomical Union can declare it to be a dwarf planet, like Pluto.

|||||||

▲ METEORITE

If an object falls to Earth from space and solid pieces survive, these pieces are called meteorites. More than 90 percent of meteorites are mostly stony; only a few are mostly metallic.

|||||||

◀ CERES

Ceres is the first minor planet to be discovered, with a diameter of about 590 miles. It's the only dwarf planet in the asteroid belt, found between the orbits of Mars and Jupiter.

|||||||

▶ METEOR

These are "shooting stars" —objects that burn up in Earth's atmosphere as they fall, leaving an ethereal streak of light. Most meteors are the size of a grain of sand.

◀ ROGUE PLANET

Theoretically, a planet can be ejected from its original solar system by a gravitational "kick" and fly through interstellar space without a sun. Astronomers haven't seen one yet, though.

▶ ASTEROID

The name means "starlike" —but asteroids are actually more planetlike, just a little small. Tumbling through space as they orbit, even small ones seem to show significant geological activity.

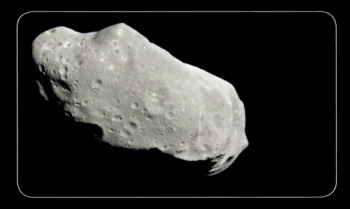

◀ COMET

These icy dirtballs of the solar system melt away as they approach the Sun, leaving tails of both dust particles and charged ions blown into beautiful arcs by the solar wind

Our solar system is littered with asteroids and comets.

Can We Catch a Ride on a Comet or Asteroid and Travel the Stars?

"It sounds like a really cool idea, because asteroids are going everywhere. But the physics prevents that from being a useful idea."

—DR. NEIL DEGRASSE TYSON

Wouldn't it be great if hitchhikers really could hop on an asteroid and tour the galaxy? Unfortunately, Neil can tell us why we can't just go ahead and do that: "It *so* would not work, because you would have to catch up with the asteroid to step onto it, and by the time you've done that, you're already in motion."

Or, as comedian Chuck Nice puts it: "It's like driving a Ferrari to catch a bus."

Exactly! "If you have the Ferrari, you don't need the bus," says Neil. "Now if you somehow found a way to get the asteroid to stop and start, but whatever power that is, if you've got it, you don't need the asteroid. You've got your own spaceship."

Riding an asteroid wouldn't be so bad: You'd have enough room to build a comfy home, mine it for its valuable minerals, and maybe even grow food on it. But you'd have to get on first—and then it'd have to escape the sun's gravity to get you to the stars. ∎

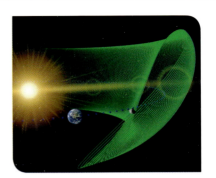

Are There Asteroids That Act Like Comets?

Some asteroids act like comets, and some comets act like asteroids. And some astrophysicists, like Dr. Amy Mainzer, will argue the night away about exactly how those objects should be classified: "It's kind of a continuum . . . We used to think that they were two totally different things, and now we know that there's a gray area in between . . . Sometimes we have asteroids who are, well, 'Sometimes we emit lots of gas and have comas that come off of us.' . . . They are basically things that can sometimes cross the boundaries between asteroids and comets." The best-known objects that behave ambiguously, like a half asteroid and half comet, are the Centaurs, a class of objects aptly named for the mythical creatures that were half man and half horse. Centaurs (the space objects) orbit primarily between the main asteroid belt and the Kuiper belt, and seem to be mixtures of ice, rock, and metal. The largest Centaurs are more than 100 miles across—way bigger than most comets. They're not rare, either; more than 300 Centaurs have been catalogued, and planetary scientists estimate there are at least 40,000 of them. ■

> *"Astronomers are going to be fighting about this for many decades to come . . . Astronomers are normally pretty quiet people, but not when it comes to this stuff."*
>
> —DR. AMY MAINZER, ASTROPHYSICIST

TOUR GUIDE

What's the Coolest Asteroid?

Asteroid 2010 TK7, discovered in 2011 with the WISE spacecraft. At least according to astrophysicist Dr. Amy Mainzer: "It's called the first-known Earth Trojan, and this is an asteroid that's stuck to the Earth in a peculiar way. It is actually trapped in a gravitational resonance with the Earth, which sounds really cool, and [it means that] Earth is following it around in its orbit around the Sun. It's kind of trapped there, and what's going to happen is, eventually, it's going to pop its way out."

DID YOU KNOW

Every 2,000 years or so, a meteoroid the size of a football field hits Earth and causes significant damage.

THINK ON THIS ▶ Why Did Comet ISON Fizzle?

"[Comets] just do whatever they feel like . . . They are exactly like teenagers, or cats . . . We just recently had a so called 'comet of the century' called comet ISON. It wasn't just a dud. It was a nothing . . . We were really hoping it would come back and make this great fireworks show, and nothing. Just like cats."
—Dr. Amy Mainzer, astrophysicist

Eureka! Asteroid Mining

Could We Fill 'Er Up on an Asteroid Someday?

One of the biggest obstacles to long-range space travel is the fuel load that must be carried into space with every launch. Turns out some kinds of asteroids have exactly the raw material needed to make rocket fuel—so could they be turned into filling stations? X Prize founder and space entrepreneur Peter Diamandis sure thinks so: "These large carbonaceous chondrite asteroids have about 20 percent water weight in them. You can extract the water, you can extract the methane, and you can break the water down into hydrogen and oxygen using the sunlight—which, by the way, is rocket fuel. A 50- to 100-meter-size asteroid . . . has more hydrogen and oxygen than was used to fuel every space shuttle launched from Earth from the beginning of the space shuttle program. So you can imagine extracting the hydrogen and oxygen and leaving it in space as fuel depots for the future missions going to the Moon and Mars." ∎

TOUR GUIDE

Could We Mine Asteroids to Fund Space Exploration?

The math is clear. A fully mined asteroid would be worth many billions of dollars. According to Neil: "An average PGM [platinum group metals] asteroid could have 30 million tons of nickel, 1.5 million tons of cobalt, 7.5 thousand tons of platinum—at current value, that platinum comes out at $150 billion." Would you have to spend that much or more, though, to mine that asteroid? And would enough investors be willing to wait long enough to get their return on this high-risk long-term investment? It'll depend on future advances in space and mining technology, as well as the vision and salesmanship of some very bold entrepreneurs.

CHUCK NICE:
So a meteor shower is basically a comet graveyard? Oh, that's awful.

NEIL:
Yeah. It's the crap left over after the Sun done did its thing.

Can I Buy and Own an Asteroid?

"If you can't have ownership, no one's going to go out there and extract the materials, and the loser is humanity."

—DR. PETER DIAMANDIS, CO-FOUNDER OF PLANETARY RESOURCES

International law about the ownership of nonterrestrial property is not yet mature. As of 2015, there exists an agreement—the Agreement Governing the Activities of States on the Moon and Other Celestial Bodies (the "Moon Treaty")—but the United States and most other spacefaring nations have not ratified it. Meanwhile, on November 25, 2015, President Barack Obama signed into law the U.S. Commercial Space Launch Competitiveness Act, under which U.S. citizens can own any asteroid resources they obtain.

"Not owning the Moon, I can agree with. But owning a 10-meter rock in space? I mean, where do you draw the line?"

—DR. PETER DIAMANDIS, CO-FOUNDER OF PLANETARY RESOURCES

At least one U.S. company—Planetary Resources— plans to launch space telescopes specifically designed to assess the commercial values of near-Earth asteroids. If their research shows that exploiting an asteroid would be profitable, they will likely chase down and capture the asteroid with rockets, claim it, and start mining it. ∎

Asteroids are particularly alluring to prospectors.

DID YOU KNOW

Metallic asteroids are often rich in palladium, platinum, rhodium, and other very valuable elements important for use in batteries, electronics, and medical technology.

THINK ON THIS ▶ Why Are Some Asteroids Metal-Rich?

Gravity works inside large planetoids, causing denser metals to sink to the center to form a metal-rich core. "Then another object smashes it to smithereens, and the smithereens are now asteroids," astro-metallurgist Dr. Neil deGrasse Tyson explains. "So now you have rocky asteroids made from the crust and mantle of that object, and then you have metallic asteroids, the precious few atoms that collected in the center. Geologists call this differentiation."

A rendering of future asteroid mining

StarTalk Live! From SF Sketchfest 2015

How Do We Find Asteroids in the Blackness of Space?

Spotting asteroids isn't just for profiting from them; it's to make sure one doesn't clobber us and cause serious destruction on Earth. The farther away we can find an asteroid, the more time we have to figure out if it's on a collision course with Earth—and maybe find a way to destroy or deflect it before it hits. Sentinal mission architect Dr. G. Scott Hubbard knows a thing or two about spotting asteroids: "It turns out if you're looking away from the sun, the sun's heating it up, and it's glowing because it's hot. So if you look in the infrared, you can find these guys."

Once you find an asteroid, you can trace its motion to determine if it'll hit or miss Earth. To find out whether it's made of stuff valuable enough to mine, you'd use spectroscopy, a tried-and-true technology. Neil explains: "This was the birth of modern astrophysics in the late 19th century. We took the spectroscope—the prism—take light, move it through the prism, out the other side. It breaks up into its component colors, like a rainbow, and in there you find embedded the fingerprints of the very chemical identity of what it is you're looking at. Bada bing!"

After that, you'd bring samples of the asteroid back to Earth for study and analysis—and for sale! "The really cool thing about this: Someone's already done it," says astrophysicist Dr. Amy Mainzer. "The Japanese sent a mission called Hayabusa, and it actually went to an asteroid, and it landed on it, and it even collected a few tiny, tiny grains off of it, and then it came back." ■

NEIL:

Gold was discovered at Sutter's Mill in 1848.

CHUCK NICE:

So why do they call them the forty-niners?

NEIL:

It took a while for that information to get out and for people to rush the state.

CHUCK NICE:

See, this is what happens when you don't have Twitter. It takes a whole year for the word to get out.

Eureka! Asteroid Mining

If Asteroid Mining Works, What Happens to My Bank Account?

When vast quantities of a scarce item become available, the law of supply and demand begins to affect its economics. If platinum becomes superplentiful, for example, its value and cost will go down. "At that price, people find more things to do with platinum that previously would not have been needed, or thought of, or dreamt of, and so now, the demand for platinum rises," explains astro-economist Dr. Neil deGrasse Tyson. "So, yes, the price per pound is less, but it's not a finite resource anymore. It's essentially unlimited."

"Technology is a scarcity-liberating force, and it's always been that way."

—DR. PETER DIAMANDIS, CO-FOUNDER OF PLANETARY RESOURCES

Centuries ago, the Spanish brought unheard-of amounts of gold and silver to Europe from the American lands they had conquered. These precious metals became so plentiful that their value plunged, causing much of Spain's vast wealth to evaporate. Today, although these metals are still used as a proxy for currency, some 15 percent of the world's gold and more than half of the world's silver goes for industrial use, not jewelry or investment. You probably have silver in your phone and gold in your teeth! ■

BACK TO BASICS

Can't We All Just Get Along?

All wars have one thing in common according to Dr. Peter Diamandis, co-founder of Planetary Resources: "Everything we fight wars over on Earth—metals, minerals, energy, real estate—those things are in near-infinite quantities in space. People look at the Earth as a very closed system, but the Earth is a crumb in a supermarket filled with resources, and if we can gain access to those resources, it uplifts everybody."

That kind of optimism pervades so many people who look boldly to the future—as long as all the benefits of progress get distributed to all who want and need them. As Benjamin Sisko says on *Star Trek: Deep Space Nine*, "You look out the window of Starfleet Headquarters and you see paradise. Well, it's easy to be a saint in paradise."

THINK ON THIS ▶ What Is Earth Worth?

"There's the oil, there's the coal, there's the minerals, which includes diamond. Then there are the elements from the periodic table that have value to our industry . . . . If I had to pick a number, I'd say a quadrillion dollars . . . The value of resources is a function of not only the demand, but the supply, and the cost of acquiring it wherever it happens to be."
—Dr. Neil deGrasse Tyson, astro-supply-and-demand-ist

PLANET EARTH

Earth is a little ball of mud orbiting a little ball of gas in a big, big universe. Oh, but what a ball of mud it is! Thanks to billions of years of continual change, our planet is the perfect place for us to have come of age in the cosmos—and now, to start our journey to the stars. What do we actually know about Earth? How should we utilize it and take care of it—and, at the same time, ourselves? The more we learn, the better we can do.

"The thought that went through my head was, 'This must be the view from heaven.' But it was replaced by a thought right after, which was, 'No, no, it's more beautiful than that. This is what heaven must look like."

—DR. MIKE MASSIMINO, ASTRONAUT

Pale Blue Dot or Big Blue Marble?

Earth is just one world among many in our solar system. Not that long ago, we humans finally began to understand where we are in space: We orbit a star, the Sun, and we're orbited by one natural satellite, the Moon. Knowledge of how Earth behaves, moves, and interacts with other bodies in the solar system has brought humans an awareness of our cosmic neighborhood and our neighbors in it.

We were built from Earth, by Earth, and we have evolved side by side with Earth for billions of years, from our first cellular ancestors to the modern junk-food muncher. When we learn about Earth's origins, we're learning about our own beginnings, and when we see natural phenomena, we build technology to study them—and learn more about ourselves in the process.

Earth is both a pale blue dot in space and a big blue marble full of beauty and majesty. It's only natural, when contemplating our planet, to think deeply and grandly about our place in the cosmos.

Set inside a marble, it's easy to see Earth in a brand-new light.

So What Is Earth, Exactly?

Can a goldfish truly understand its bowl without taking a look from the outside? We use our eyes and our machines to look to the stars, but when we peer back at ourselves, this is what we see.

||||||

◄ "BLUE MARBLE"

The original "Blue Marble" photograph was taken from Apollo 17 on December 7, 1972, at a distance of 28,000 miles from Earth.

||||||

▶ EARTH'S HORIZON

An Expedition 7 crew member on board the ISS in July 2003 captured this view of Earth's horizon as the sun set over the Pacific Ocean.

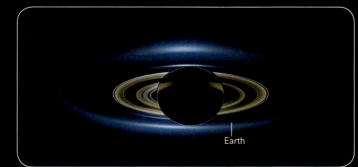

Earth

||||||

◄ "THE DAY THE EARTH SMILED"

Earth is in the background of a photograph of Saturn, taken from Cassini on July 19, 2013, at a distance of 900 million miles.

||||||||
▶ "EARTHRISE"

Declared "the most influential environmental photo ever taken," "Earthrise" was shot from Apollo 8 on December 24, 1968, from a distance of 240,000 miles.

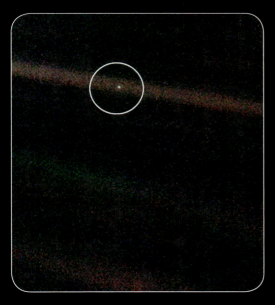

||||||||
◀ "PALE BLUE DOT"

Transmitted from Voyager 1, the "Pale Blue Dot" photograph of Earth was shot on February 14, 1990, at a distance of 3.76 billion miles.

||||||||
▶ EARTH AT NIGHT

Perhaps more than any other images, pictures like these confirm the presence of humanity on our planet. Extraterrestrial visitors, seeing the light we make in the dark, would have no doubt we are here.

The famous "Earthrise" picture was taken by astronauts on Apollo 8.

Space Chronicles (Part 1)

How Did Apollo 8 and "Earthrise" Influence the Environmental Movement?

"We went to the moon, looking to discover it, and we looked back and we discovered Earth for the first time."

—DR. NEIL DEGRASSE TYSON

Today, we hardly think anything of the pictures of Earth taken from space. After all, you can see them on every TV weather report. Before the Apollo era, though, an image like that was unthinkable.

On Christmas Eve in 1968, astronaut William Anders and his Apollo 8 crewmates held a historic live-from-space broadcast that showed pictures of Earth and the Moon viewed from their spacecraft. Upon their return, "Earthrise" (above) became one of the most iconic images in history. For the first time, the people of the world were able to appreciate Earth as a whole, from afar—beautiful, isolated, and fragile.

The environmental movement in the United States was young then, having started in earnest in 1962 with marine biologist Rachel Carson's book *Silent Spring*, which warned of the dangers of pesticide overuse. But "Earthrise" cemented the realization that our planet's environment needed help. Two years after Apollo 8, on December 2, 1970, President Richard Nixon approved the establishment of the Environmental Protection Agency. ■

The Day the Earth Smiled

Two decades after "Pale Blue Dot," that iconic Voyager 1 picture of Earth, was taken, planetary scientist Carolyn Porco was the leader of the imaging team for Cassini-Huygens, a multiyear mission to Saturn and its moons. Why not get a new portrait of our planet? she thought. Here's how she recalls the day: "The 15 minutes that it was happening and I'm looking where Saturn is; and I'm thinking, Wow, there's a camera there taking our picture; and knowing that people all over the world were doing the same thing—it was fabulous."

Her plan was to tell everyone on Earth what was happening, to be ready for it, and to say "Cheese!" for the spacecraft's camera. On July 19, 2013, that's exactly what happened. "And I'm thinking, Why don't we tell people in advance: 'Your picture is going to be taken from the outer solar system, from a billion miles away' . . . How better to let them know how far humans have come in the exploration of the solar system," she recalls. "It becomes something personal to them . . . because the idea was to smile in celebration. To get this communal feeling out of people, this kind of cosmic love . . . And one person wrote, 'You know, darn it, we may be floating around on a dust mote, we may be transient, but for 15 minutes, we were there, we were aware, and we smiled.' " ∎

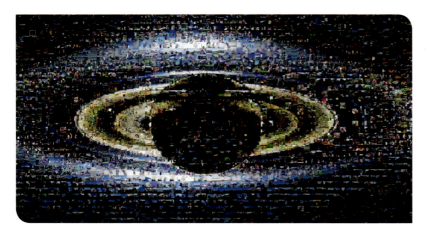

"The Day the Earth Smiled" is re-created from a collage of 1,600 images submitted to NASA's "Wave at Saturn" campaign.

BIOGRAPHY

Carl Sagan and the Spirituality of Space

Carl Sagan (1934–1996) was the foremost science communicator of the 20th century—but not just because he was a top scientist who could explain his research clearly. Sagan helped people connect with science in every context; whether it was technical, practical, hypothetical, or even spiritual, he showed everyone how the universe related to them. Sagan was Jewish, but his sense of the cosmic transcended any single faith tradition. As he wrote in his 1994 book, *Pale Blue Dot*, about the tiny image of Earth from Voyager 1: "The aggregate of our joy and suffering, thousands of confident religions, ideologies and economic doctrines . . . every teacher of morals, every corrupt politician, every 'superstar,' every 'supreme leader,' every saint and sinner in the history of our species, lived there on a mote of dust suspended in a sunbeam."

Cosmic Queries: Planet Earth

How Was Earth Formed?

About 4.6 billion years ago, soon after nuclear fusion in the Sun began, the leftover material from the Sun's formation settled into a thin disk rotating around our infant star. This disk contained enough bits of rock, metal, dust, and gas to make hundreds of Earths; and over millions of years, in a process called accretion, the bits came together to build pebbles, then rocks, then boulders, then planetesimals, then planetoids—and then, finally, planets.

NEIL: *If you don't have enough gravity to begin with, you hold your rocks; you don't hold your gas.*

CHUCK NICE: *You mean Uranus is gassy?*

Earth formed in a region of the solar system where there wasn't a whole lot of material, and what was there consisted mostly of rock and metal. That's why our planet is small, solid, and rocky-metallic—a terrestrial planet like Mercury, Venus, and Mars. Jupiter, Saturn, Uranus, and Neptune, on the other hand, formed where there was much more planet-building raw material, but it was mostly gas and almost no metal, and so they became gas giant planets with cores of rock and ice.

This simple picture of planetary formation is hardly the whole scientific story, of course. Lots of mysteries remain. For example: Zipping around the Sun at thousands of miles an hour, the original bits of matter probably bounced off one another, so how did they join to form larger bodies? There is plenty left to learn! ■

BACK TO BASICS

What Was the Greatest Wrong Idea Ever?

"There's nothing wrong with a wrong idea," says Dr. Neil deGrasse Tyson, astro-wrong-icist. "[As long as] it helps you try to solve a problem. One of the greatest wrong ideas ever was the geocentric universe, with the Earth in the middle."

The geocentric model of the universe was proven to be spectacularly incorrect—yet it helped explain how things move in the universe as viewed from Earth. Even more importantly, the people who codified geocentrism—Aristotle, Ptolemy, and other great thinkers—created a model that could be tested scientifically, and that allowed their descendants to keep improving humanity's view of the cosmos.

THINK ON THIS ▶ How Did Our Sun Form?

Nearly five billion years ago, a dense core of gas collected in a vast interstellar cloud trillions of miles across. That cloud collapsed in on itself, creating a core with a temperature of more than 20,000,000°F and pressure two billion times that of Earth's atmosphere—nearly 30 billion pounds per square inch—conditions extreme enough to induce hydrogen to fuse into helium. Then the collapse halted, and the Sun was born.

Where Do Earth's Heavy Metals Come From?

You don't get the real answer to Neil's question until you study astrophysics and ask again, Where do these elements come from? "They are cooked in the crucibles of high-mass stars, forged from small elements like hydrogen and helium; they are fused together to make high-mass elements, and they ride their way up the periodic table of elements," explains astro-chemi-cist Dr. Neil deGrasse Tyson. "And then that same star explodes, scattering its guts into the galaxy, out of which you make subsequent solar systems."

The astrophysicist's answer to this question shows how we're connected to the entire universe at literally the atomic level. When stars much more massive than our Sun explode as supernovae, atomic nuclei and subatomic particles crash into one another in super-high-energy interactions, and heavy elements are created in a fraction of a second. This is "rapid-process nucleosynthesis," which produces most of the heavy elements in the universe.

Stars that don't explode, like our Sun, can also produce heavy elements during a short period of their lifetimes, when they swell in size to become giants. This is "slow-process nucleosynthesis," which takes thousands of years. Some elements can only be made by one process, while others can be produced by either. ∎

"When I was a kid in chemistry class, I'd ask the chemistry teacher, 'Where do these elements come from, that sit up there on the periodic table?' 'We find them in the ground.' That was the chemistry teacher's answer."

—DR. NEIL DEGRASSE TYSON, ASTRO-TEEN-ICIST

A marble statue of Atlas, the Greek god of astronomy and navigation, bearing the weight of Earth on his shoulders

YouTube: Neil deGrasse Tyson on Lunar Phases and Tides

Does the Full Moon Make Earth's Tides Stronger?

TOUR GUIDE

The Moon is the main reason why oceans have tides: The Moon's gravity has a stronger influence on the side of Earth nearest to it, and it draws Earth's water toward it ever so slightly, which causes daily tides. Maybe you have noticed that tides run higher during a full moon—but it's not the Moon's phase that causes that. The Moon's tidal effects depend only on its mass and distance from Earth, which stay almost constant no matter what its phase.

"Why do we have a higher tide during full moon? Because the Sun's tides add to the Moon's tides. It's the Sun. Blame the Sun."

—DR. NEIL DEGRASSE TYSON, ASTRO-SUN-ICIST

So what makes high tides get even higher? It has to do with the Sun. The Sun also has a gravitational pull on Earth, although a little less than half that of the Moon. A full moon happens when the Sun lines up with Earth and the Moon, and in that position, the Sun adds its gravitational influence on our oceans. Those higher-than-high tides are called spring tides, because the water seems to "spring" upward just a little more than usual. Between the full and new moon, during the quarter phases, the Sun pulls at right angles to the Moon's pull, creating a weaker result and relatively lower high tides, called neap tides. ∎

What's the Deal With the "Super Moon"?

Now and then, the buzz goes around: Watch for the super moon tonight. So what's a super moon? Don't get astro-moon-icist Dr. Neil deGrasse Tyson started: "I don't know who first called it a super moon. But if you have a 16-inch pizza, would you call it a super pizza compared to a 15-inch pizza? The Moon's orbit around the Earth is not a perfect circle. Sometimes it's closer; sometimes it's farther away. And every month, there's a moment when it's closest. Occasionally, that moment coincides with a full moon. People are calling that a super moon. But there are super half-moons. Every month, every one of the phases is the closest. I don't hear people say, 'Oh, cool, super crescent.' I told you, don't get me started."

THINK ON THIS ▸ Why Is the Moon Short on Iron?

A much lower percentage of the Moon's mass is iron compared with Earth's, yet the Moon and Earth have nearly identical crusts. Astronomers think that's because a Mars-size planet crashed into Earth billions of years ago, throwing a lot of rock into orbit, but not a lot of metal. That rocky material eventually formed the Moon.

Meltwater pools on the Greenland Ice Sheet may influence Earth's rotation.

How Do Earthquakes and Melting Glaciers Change Earth's Rotation?

HOW MUCH DOES EARTH WEIGH?

"You could justifiably say the Earth is weightless in space," says Neil. Its mass, on the other hand, is about six billion trillion tons.

Rotating bodies, be they planets or people, follow a fundamental law called conservation of angular momentum. As Neil explains it: "Skaters do this when they pull their arms in—they are changing their own moment of inertia. What happens to their rotation rate? They go faster. And how do they stop? They put their arms back out, and they can stop on a dime. So they change their own rotation rate by bringing their arms in or out."

The December 2004 earthquake in the Indian Ocean, the third strongest ever recorded, shortened the length of our day by an estimated two microseconds. Melting glaciers, on the other hand, increase Earth's diameter by creating higher sea levels, lengthening the day by about a thousandth of a second since the year 1900. Think of the earthquake as the skater pulling in her arms; think of Earth with higher sea levels as the skater spreading them out. "You can calculate how much we change after every earthquake, because an earthquake is a redistribution of the continental shelf," says Neil. "In fact, melting glaciers change the rotation rate of the Earth, too." ■

Cosmic Queries: Grab Bag

Why Do Planets Revolve Around the Sun on the Same Plane?

DID YOU KNOW
The solar system extends for billions of miles outward from the Sun, yet it contains less than one-tenth of one percent of the Sun's mass—and most of that is the planet Jupiter.

Over the course of a year, the path of the Sun across the sky traces out a surface called the ecliptic plane. All the planets move on or very close to this plane as they orbit the Sun—and they all do so traveling in the same direction. Amazing coincidence? No, there's a reason. Let Dr. Neil deGrasse Tyson, astro-pancake-ist, explain: "In the formation of the solar system, you have this huge rotating gas cloud, and it wants to collapse under its own gravity. As it collapses, it pancakes, and out of that revolving material, it makes objects. And so if you make planets out of that rotating material [as it's moving], those planets all orbit the sun in the same direction, on the [same] plane . . . Everything's in a plane, and they're going in the same direction together." ∎

> "If you come upon a solar system, and a planet is orbiting some other funky way, you can bet your . . . orbital parameters that planet was captured from another place outside your solar system."
>
> —DR. NEIL DEGRASSE TYSON

A rendering of the eight planets of our solar system on the same plane

DID YOU KNOW
According to Neil, if Earth's rotation speed suddenly doubled, the force required to make that happen would "flatten you into a pile of goo."

THINK ON THIS ▶ Why Are Days Shorter at the Winter Solstice?

Earth's polar axis slants. So as Earth orbits the Sun, sometimes the Northern Hemisphere leans more toward the Sun and sometimes it leans away. Daylight hours can vary throughout the year as a result. On the winter solstice, the North Pole is leaning farthest away from the Sun, so daylight hours are shortest—but simultaneously, the Southern Hemisphere is leaning as close toward the Sun as it ever gets, and experiences the longest day of the year.

What Happened to the Meteors?

Huge dents in Earth's crust have been made over the eons by meteors slamming into our planet. Depending on where they hit and the environmental conditions surrounding them, those craters are still visible millions of years after impact. But the rocks themselves are gone! Why?

Scientists have used computer simulations and calculations to envision such an event and solve the puzzle. When a meteor of great size strikes a solid surface at hypersonic speed, the object explodes on impact. The detonation creates a perfectly symmetrical round hole, no matter what angle the collision occurred—and there's nothing left of the meteor but vapor. ■

BACK TO BASICS

How Thick Is Earth's Atmosphere?

Gas molecules in Earth's atmosphere extend thousands of miles above sea level. Above a couple of hundred miles, though, the atmosphere is so thin that we can hardly distinguish it from the vacuum of space. So exactly how do we measure the "thickness" of Earth's—or any planet's—atmosphere? One way is to measure the pressure at the bottom of the atmosphere. On Earth, that's about 14.7 pounds of force per square inch. That means typical human adults constantly experience nearly 20 tons of force on their bodies, just from Earth's atmosphere! As impressive as that sounds, Venus has an atmosphere 90 times denser than Earth's—and the atmospheres of the gas giant planets Jupiter, Saturn, Uranus, and Neptune are thousands of times denser than that.

Meteor Crater
A 150-foot-wide metallic meteor struck the Mogollon Rim in modern-day Arizona 50,000 years ago, leaving a mile-wide hole that's 60 stories deep.

Chesapeake Bay
A mile-wide meteor hit the Eastern Shore of the United States 35 million years ago, pushing the surrounding land downward. That indentation eventually filled with water.

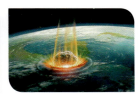

Chicxulub
About 65 million years ago, a 10-mile-wide meteor pounded the planet, off the coast of modern-day Mexico's Yucatán Peninsula. Its impact on Earth's atmosphere may have led to the extinction of the dinosaurs.

"Meteor Crater was purchased by a metal speculator, because he presumed that the huge object that made the crater was still buried underneath . . . Didn't find a damn thing, poor guy."

—DR. NEIL DEGRASSE TYSON, ASTRO-CRATER-CIST

THINK ON THIS ▸ How Does Light Get Through the Greenhouse Effect?

Greenhouse gases keep heat from escaping the atmosphere—why don't they keep sunlight from coming in? Doesn't sunlight heat Earth? "The Sun does not heat the air," explains astro-solar-cist Dr. Neil deGrasse Tyson. The Sun heats the ground, and the ground heats the air. Visible light hits Earth and heats it. Earth reradiates that same energy as infrared, and it's the infrared that gets trapped by the greenhouse gases."

Carved channels on Mars may indicate the planet once had surface water.

What Can We Learn About Earth From Venus and Mars?

Was Venus once Earthlike?

Venus and Mars have undergone climate changes that dwarf any earthly scale—and have mighty stories to tell. Says astrobiologist Dr. David Grinspoon about Venus: "Venus started off Earthlike. As near as we can tell, it did have oceans and water and was cooler when it was young, and then as the Sun heated up . . . it passed a point where things got hot enough, so the oceans started to evaporate. As the oceans evaporated, there was more water vapor in the air. Water vapor is a strong greenhouse gas, so water vapor in the air makes it hotter at the surface, which leads to more evaporation of the oceans, more greenhouse gas in the atmosphere. It's a positive feedback—it runs away. It got to the point where the oceans literally boiled off."

Mars experienced a greenhouse effect that didn't run away—it ran out. Billions of years ago, Mars probably had liquid water on its surface, but things got too cold, and the atmosphere dissipated into space. Martian oceans either evaporated or froze solid deep underground. ∎

YouTube: Neil deGrasse Tyson Explains the Aurora Borealis

What Causes an Aurora Borealis?

In addition to light, electrically charged subatomic particles stream outward from the sun into the solar system. Given time, these particles would damage living tissue and render Earth uninhabitable. Fortunately for us, Earth generates a magnetic field as it spins, and this field directs most of these charged particles away from Earth's surface and harmlessly into space.

Sometimes, the number of charged particles spikes upward—for example, if the Sun has a solar flare or a coronal mass ejection. The extra particle flux streams along Earth's magnetic field lines toward the magnetic pole, crashing into the molecules of our upper atmosphere. The resulting release of energy is the cosmic light show known as the northern lights, or the aurora borealis. When it happens in the Southern Hemisphere, it's called—you guessed it—the southern lights, or the aurora australis.

Auroras are especially beautiful when viewed from space—where they appear as flowing rivers of eerie light cascading downward toward Earth's surface. We see them, sometimes spectacularly, around other planets, too. ■

A vivid aurora borealis over Russia

BACK TO BASICS

Why Do Stars Twinkle?

Even on the calmest days, the air in Earth's atmosphere jiggles and shimmies. Twinkling happens when the light from a distant point travels through that atmosphere; its light beam dances ever so slightly in random directions many times—sometimes thousands of times—each second. To an astronomer on Earth, looking at a distant star through the atmosphere is like looking up at a firefly from the bottom of a swimming pool. That's why telescopes in space are so important—they rise above the obscuring effects of the atmosphere and send down images with clarity unobtainable on terra firma.

"'Twinkle, twinkle, little star, how I wonder what you are' is the bane of all modern astronomy, because the twinkling is atmospheric."

—DR. NEIL DEGRASSE TYSON,
ASTRO-MUSICOLOGIST

The Past, Present & Future of Space Telescopes?

From Galileo's handheld device to today's orbiting marvels, telescopes revolutionize our understanding of who we are. "Every time you turn a bigger telescope to the night sky, we end up smaller than we had previously imagined . . . It is an ego-dismantling device," says Neil.

|||||||

◄ GALILEO'S TELESCOPE

This wooden tube containing two lenses gave Galileo Galilei barely 20 times the light-gathering power of unaided eyes—and revealed that Earth was just a planet in a whirling Sun-centered solar system.

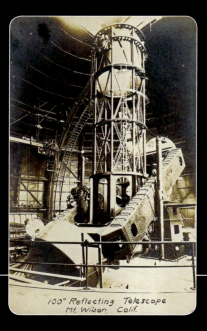

100" Reflecting Telescope Mt. Wilson, Calif.

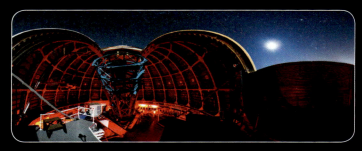

|||||||

◄ HOOKER TELESCOPE

With this telescope, located at Mount Wilson Observatory (above) in California and powered by a mirror 100 inches across, Edwin Hubble discovered that ours is but one galaxy among billions in the universe.

|||||||

◄ HALE TELESCOPE

By many measures the most productive telescope in history, this iconic 200-inch telescope on Mount Palomar in California brought astronomy into the modern era of observational cosmology.

IIIIIII
▶ KECK TELESCOPES

Sitting 14,000 feet above sea level at an observatory on Mauna Kea in Hawaii, the twin Keck I and Keck II telescopes each combine 36 hexagonal mirrors into a single light-gathering surface nearly 400 inches across.

IIIIIII
▲ ARECIBO OBSERVATORY

This radio telescope, 1,000 feet across and larger than 25 football fields, was built into a natural valley in Puerto Rico. Its many uses include listening for signs of extraterrestrial intelligence.

IIIIIII
◀ HUBBLE SPACE TELESCOPE

The most important scientific instrument of its generation, Hubble has been upgraded five times by astronauts tending it from orbiting space vehicles. Today it's about 100 times more powerful than when it was launched a quarter century ago.

IIIIIII
▲ CHANDRA X-RAY OBSERVATORY

CXO is sensitive to light that would never reach Earth's surface and that human eyes can't detect. With optics specially positioned to catch x-rays, it almost looks as if it's pointed backward.

IIIIII
▼ JAMES WEBB SPACE TELESCOPE

The scientific successor to the Hubble and Spitzer Space Telescopes, JWST will be sensitive to both optical and infrared light. Its adaptive optics will be able to pick out infant galaxies as well as planets orbiting distant stars.

IIIIIII
▶ KEPLER SPACE TELESCOPE

Named after the astronomer who discovered the elliptical orbits of planets, this special-purpose space telescope has been used to find thousands of planets so far—all of them orbiting stars other than the Sun.

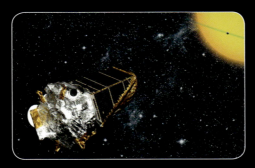

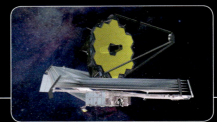

"So you want to talk about spirituality, you take a swig of water . . . It contains water molecules that have passed through the kidneys of Abraham Lincoln, of Genghis Khan, of Jesus . . . "

—DR. NEIL DEGRASSE TYSON

CHAPTER TWO

How Do We Get Water?

Water is an amazing substance—yet it's so ubiquitous, we almost take it for granted. It's just two hydrogen atoms attached to an oxygen atom, forming a wide V-shaped structure. That shape allows water to dissolve more substances than almost any other known liquid and yet stay stable for eons without denaturing or dissipating.

Water's such a good solvent, though, that impurities, microbes, or even poisons can easily get into it as it travels. So when we get water for daily use, we sometimes have to bring it from far-away places to be sure it's clean. Constant vigilance to protect our water supplies is a must for our well-being—but with a swelling population and ever intensifying competition for living space and resources, that's getting harder and harder.

What if we don't have enough water? Or what if the water is plentiful but it's dirty or salty? Can we get more from the cosmos? After all, that's where all of our water came from, a long time ago.

The abundant water on our planet makes Earth unique.

Pilgrims bathe in the Narmada's 160-foot-tall Kadil Dhara waterfall.

Where Does Our Water Come From?

"Early on, when the solar system was first forming, we do think there was a period when the Earth pretty much got pelted by asteroids and comets, and that could have been when the water got here."

—DR. AMY MAINZER,
ASTROPHYSICIST

Most of Earth's water is locked in rock. You can't tell from touching it, for example, but up to a quarter of the volume of a block of sandstone could be water. For some volcanic rock and limestone, that fraction can go up to half. The amount of water in Earth's outermost layers—the crust and upper mantle—is estimated to be more than 10 times the total on Earth's surface. Early in Earth's geological history, over hundreds of millions of years, water was carried to the surface, most likely by volcanoes, and produced our oceans, lakes, and rivers.

So how did the water get there in the first place? For that, we have to look to the universe—because that's where all of Earth's building blocks came from, billions of years ago. ∎

Where Have All the Comets Gone?

"Not all water is created equal."
—DR. NEIL DEGRASSE TYSON

I f the water molecules on Earth arrived here in comets, those original icy dirtballs melted long ago. But there are still comets in our solar system that can help us test this hypothesis, by comparing the atomic properties of today's water with the properties of today's comets.

So did water arrive in comets? Maybe, maybe not says astro-ocean-ist Dr. Neil deGrasse Tyson: "We checked some comets, and we noticed the water in those comets doesn't match the water in our oceans, and that got us worried . . . But then we did find a couple of comets that did match the oceans. The jury is still out, just so you know, [about] the origin of Earth's water. We know at least some came from volcanoes, and we're pretty sure there's a class of comet that did not supply the water, and other classes of comets that might."

Another point to consider: About one-half of one percent of Earth's mass is water, and about one-tenth of that water is in Earth's oceans. Just to fill the oceans, at least one billion good-size comets would be required. So there's a lot of our planet's water supply that remains unexplored, untested, and unexplained. ■

Some of our ocean water may have come from comets.

DID YOU KNOW

If the Chicxulub impactor, which wiped out the dinosaurs 65 million years ago, was indeed a comet, it would have deposited enough water on Earth's surface to flood the state of Florida to a depth of 10 feet.

THINK ON THIS ▶ Are Comet Water and Asteroid Water Different?

"Water is not in asteroids the way you find water in a comet. A comet [has] huge chunks of ice, and you can just carve out a piece and there it is. [In an asteroid,] water is sort of blended in with the minerals, so you actually have to extract it. You actually have to mine the water."
—Dr. Neil deGrasse Tyson

"It turns out that the Moon wasn't this dry body in the sky that we thought it was, because the Moon and Earth share a history."

—DR. YVONNE PENDLETON, DIRECTOR OF NASA'S SSERVI

StarTalk Live! From SF Sketchfest 2015

What's Wetter: Europa, Mars, or the Moon?

Jupiter's moon Europa has a crust made of ice that looks almost exactly like Earth's polar ice cap. That's led astronomers to conclude there is a lot of water deep under Europa's surface. We have yet to see one drop up top, though.

On the other hand, it's been proven that Mars has oceans of ice beneath its surface—and once in a while, that ice even appears as liquid water on the Martian surface. Warm weather can cause water to ooze out of Martian mountain walls and run downhill until it evaporates, leaving trails of mineral deposits for our orbiting telescopes to see.

It looks like there may be some water on the Moon, too, possibly in the form of ice preserved in shady craters, or chemically bound in rocky material blasted into space from a water-rich Earth four billion years ago. "Apollo astronauts picked up rock samples, and later we were able to find out there was quite a bit of water in those rocks," says Dr. Yvonne Pendleton, director of NASA's SSERVI. "The Earth was probably wet at the time when that big Mars-size object hit the Earth and created the Moon. Some of that water went to the Moon." ■

"By the way, if you're a comet and you hit the Earth, all of your water vaporizes on contact so you would become steam, and you would have to condense back out later."

—DR. NEIL DEGRASSE TYSON, ASTRO-STEAM-ICIST

What Are Water's Secrets?

magine dropping an ice cube into a steaming cup of hot tea. There you have water in all three phases—solid (ice), liquid (tea), and vapor (steam)—each in the vicinity of one another. The surfaces between those phases interact in weird ways that still aren't fully understood scientifically.

When you slip on ice, what's happening? The physics of surfaces—solid or liquid, sandpaper or ice—is surprisingly complex. What we do know is that ice creates very little friction when its surface touches another solid surface. One

"Frozen water is less dense than liquid water . . . and ice floats. In the winter, the top surface [of a lake] gets cold. The top water will freeze, and not drop, thereby insulating the liquid water below, protecting the fishes through the winter months. It's a remarkable feature of water."

—DR. NEIL DEGRASSE TYSON, ASTRO-ANGLER-IST

common misconception is that ice skates melt the ice they run over, but no: That slipperiness is already a function of the low friction of the ice.

Why is ocean water salty? Salt starts as a chemical component of rock. Water running over rocks can dissolve the chemicals in those rocks. When all that water collects in one place, the water can evaporate away, but the salt stays behind. That happens in the oceans, of course, but also in some inland bodies of water, known for their high salinity, such as the Dead Sea and the Great Salt Lake. ■

Its unique properties make frozen water—ice—dangerously slippery.

What Is the Water Cycle?

Arriving on Earth from space billions of years ago, then bubbling up onto Earth's surface through fissures and volcanoes, H_2O molecules do a perpetual dance of transformation that makes life possible.

‖‖‖‖‖
◀ EVAPORATION

Liquid water or solid ice warmed by the sun or the ground below absorbs energy and transforms into gaseous vapor. Floating into the sky, it can stream freely or collect into denser concentrations.

‖‖‖‖‖
▶ CONDENSATION

As the vapor hits chillier air, it loses heat energy and transforms back into water droplets and ice crystals. Clouds look serene from the outside, but inside there's constant swirling motion.

‖‖‖‖‖
◀ PRECIPITATION

Little bits of water and ice glom together with each other and onto bits of dust, soot, or other small particles to form raindrops and snowflakes. When they get too heavy, they fall to Earth.

|||||||
▶ INFILTRATION

Rain and liquefied snow seep into the ground through porous rock or soil. Groundwater collects. Plants suck water through their roots to use in biological processes. Animals drink.

|||||||
◀ FLOW

If too much water falls to be absorbed into the ground, the excess collects in puddles and pools. Gravity makes the water move downward toward Earth's center, creating runoff, then streams, rivers, lakes, seas, and oceans.

|||||||
▲ RETURN

Plants transpire. Animals perspire. (And, yeah, they also pee.) One particular animal—humans—uses water for everything from agriculture to industry to entertainment, and then dumps it back out into the environment. The water cycle repeats.

Let's Get Spiritual

You've heard of holy water? Actor and comedian Jason Sudeikis thinks the water cycle proves the point. Not only does water exist in three phases, he says, but it also exists on Earth and in the heavens above: "What more of a spiritual journey exists than existing on Earth, like water does—ascending to the heavens, hanging out there for a few days, then coming back down and then helping people out again? . . . Like, the water cycle is the Trinity, you know. Three's [a] magic number." ■

StarTalk Live!: Water World (Part 1)

Can I Drink Heavy Water?

Ordinary water has two hydrogen atoms and one oxygen atom: H_2O. The stuff we call heavy water has two deuterium atoms and one oxygen, and it's a little bit denser than ordinary water. Any water could have a tiny bit of heavy water in it, but to get a glass of it, you would have to start with a huge amount of water and extract the heavy water out.

Heavy water got a bad rap because of its use in developing nuclear weapons. Could you still drink it? Sure, says Neil, "You would slow down, though."

Comedian Eugene Mirman isn't buying it and asks, "Really? You wouldn't die?"

Not at all. Heavy water is made with deuterium atoms, which are actually hydrogen atoms with an extra neutron in each nucleus. So heavy water is practically identical to regular H_2O. It's not poisonous or anything. But it might not behave exactly the way you expect water to when you drink it.

Lab mice have had the experience, but we can only speculate what would happen to a human drinking heavy water. Judging by the numbers, a molecule of heavy water is 11 percent more massive than a molecule of regular water, and two-thirds of your body mass is water—so after drinking heavy water, you would be just a little heavier. ■

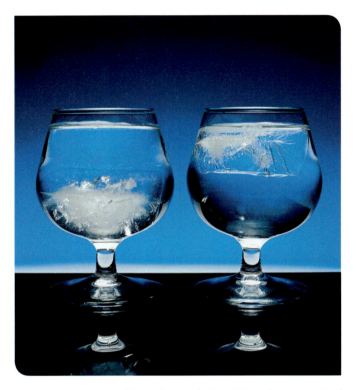

Two glasses demonstrate the concept of heavy water and regular water.

DID YOU KNOW
In every million gallons of seawater, there is about a double-shot glass full of naturally occurring heavy water.

THINK ON THIS ▶ How Much of Earth's Water Is Drinkable?

OK, so about 0.02 percent of Earth's mass is surface water. How much of that can I drink?

DR. TESS RUSSO, GEOSCIENTIST: "About 3 percent of the water on Earth is fresh water, but most of that —two-thirds of it—is locked up in glaciers and ice."

NEIL: "So we should melt the glaciers and get more."

DR. RUSSO: "We are."

StarTalk Live!: Water World (Part 2)

What Is the Future of Our Water?

Humans use lots and lots of water to run society and civilization. As more and more of humanity congregates and resides in and around cities, delivering water to them requires ever greater effort—and ever larger sources of clean, fresh water to tap.

New York City, for example, is the most populous city in the United States, and its residents use about 1.5 billion gallons of water each day. Its supply of fresh water comes from three reservoir systems far to the north, covering about 2,000 square miles of watershed. Those northern regions, which are already home to hundreds of thousands of people, are now undergoing significant development and construction, potentially endangering the city's water supply. Furthermore, the tunnels that carry this water into the city are about a hundred years old. If they fail, millions of people will be in big trouble. ■

An illustration shows a greener world.

TOUR GUIDE

Why Was Our Holy Lake Killing Us?

In the Himalaya, there is a village with a beautiful lake that has for generations had a spiritual purpose. Buddhist spiritual leader His Holiness the Gyalwang Drukpa, explains what happened at this lake: "When people die, the dead body goes in there, and horse goes in there, and the dead yak, dead sheep—they're all dumped in there. And I have noticed that so many people are sick, especially [with] skin disease. And every doctor [that] goes there, they can't help . . . I asked them to clean the lake, to clean the river, and I asked them to not throw the dead body there . . . And within two years, the whole entire disease stopped."

"Instead of investing in huge highways and subsidies that go out into settling the landscapes around our cities, we should be investing in police protection, great schools, and great health care in the cities, making people want to live here, and then saving the landscapes to filter water."

—ROBERT F. KENNEDY, JR., ATTORNEY AND PRESIDENT OF WATERKEEPER ALLIANCE

Where Do We Use Water Most?

Clean water plays a huge role in modern life: drinking, bathing, housekeeping, and watering the yard and garden. The EPA estimates the average U.S. household uses 300 gallons a day. With roughly 125 million households in the country, that means we're using more than

"There are [already] a lot of technologies out there right now that if widely used would dramatically reduce the amount of water that's needed."

—ROBERT F. KENNEDY, JR., ATTORNEY AND PRESIDENT OF WATERKEEPER ALLIANCE

35 *billion* gallons of water a day. But that's just the beginning. Households in the U.S. represent less than 10 percent of water usage. Agricultural irrigation systems and power plants each use more than four times the water that households do. Here are some of the major water users. ■

DID YOU KNOW?

According to UNESCO, nearly one out of every nine people around the world does not have access to clean drinking water, and at least one-third of the world's population does not have the water necessary for adequate sanitation. As the world population bulges beyond seven billion, these demands become greater. At the same time, climate change means a warmer planet, and many areas—including parts of North America—face serious drought conditions.

||||||
▼ IRRIGATION

Large-scale farming requires steady irrigation. Estimates put this use across the United States at **128 billion gallons** daily.

|||||||
▲ TEXTILE PRODUCTION

It takes **700 gallons** of water to sew, size, and dye every T-shirt manufactured in the United States.

|||||||
◄ POWER GENERATION

Steam-driven turbine generators, which create thermoelectric power for households, use about **160 billion gallons** of water every day.

|||||||
▲ HOUSEHOLD LEAKS

Every day, leaking pipes and faucets in U.S. household water systems lose at least **3 billion gallons**—literally down the drain.

|||||||
▲ FRACKING

In eastern Pennsylvania alone, about **30 million gallons** of water are used daily to operate hydraulic fracturing wells—"fracking" operations—to extract natural gas.

|||||||
▶ BOTTLED WATER

About **10 billion gallons** of water are bottled in the U.S. annually—that's about 150 bottles per person a year—and the number is growing rapidly.

Will World War III Be About Water?

DID YOU KNOW
In sub-Saharan Africa, most people gather their water from wells, springs, lakes, and rivers. The time spent gathering water there totals 40 billion hours a year.

Millions of people worldwide are forced to leave their homes each year because of water—either too much of it causing flooding, or too little of it causing drought and famine. "The Pentagon has done two assessments over the last decade," says Robert F. Kennedy, Jr., attorney and president of Waterkeeper Alliance, "and in both those assessments they've said that global warming, and particularly water shortages, are the principal threat to America's national security because of the disruption to global political systems and populations that it's going to cause."

High-stakes business investments come into play as well, continues Kennedy. "In the last 10 years, water privatization has become a trillion-dollar industry, according to the World Bank. We've already seen water wars fought all over the world—in Bolivia, in Cochabamba; and Belize; and many other countries where foreign companies have come in, privatized local water supplies, and then raised rates and literally killed poor people who can't afford those higher water rates."

Displaced and desperate people are much more likely to be part of civil and political strife, including crime and terrorism. Furthermore, many of the most politically sensitive places in the world, such as the Middle East, are in desert regions. Water issues can fuel violent conflict. ■

If only water wars could be fought with water guns.

"There are millions of environmental refugees every year, throughout the world, who are driven off their land and who are creating civil strife, and political strife, and security problems that affect the United States because of water shortages."

—ROBERT F. KENNEDY, JR.,
ATTORNEY AND PRESIDENT OF WATERKEEPER ALLIANCE

THINK ON THIS ▶ How Much Does That Bottle of Water Really Cost?

About 50 billion plastic bottles of water are sold in the United States each year. It takes almost 20 million barrels of oil to make just the bottles, enough to keep a million cars on the road all year, or to provide electrical power to 200,000 homes. Three-fourths of these bottles aren't recycled, creating more than a billion pounds of plastic trash every single year.

"New York City's tap water is so good because of how thriving the estuary of the Hudson River Valley is—and in San Francisco we've got a pretty good one [too]."

—ADAM SAVAGE, COHOST OF *MYTHBUSTERS*

StarTalk Live!: Water World (Part 1)

What's Going On in California?

A historic multiyear drought and heat wave, combined with intensive use of water, has left the state of California and its nearly 40 million residents in a water crisis. Add to it that 2014 and 2015 were the two warmest years in the state's 121 years of record keeping. Future rain and snow will need to make up what has now become a shortfall of 11 *trillion* gallons of water.

It's not just a matter of rainfall and snowfall. Longstanding archaic governmental policies—in California and its neighboring states—have deeply affected the water situation there. Governmentally decreed water allocations more than a century old still hold sway and create some of the problems. Attorney and president of Waterkeeper Alliance Robert F. Kennedy, Jr., explains: "The governments wanted to get white men to move out into the western states, to settle the states, to take [land] away from the Mexicans and the Indians. So what they did is, they said, 'If you come out here, you can own the water—and as much as you can use, you own it.'

"And so there are very, very irrational laws out West that incentivize people to use as much water as they can, to grow rice in the desert, to grow alfalfa in the desert, and to grow cities like Las Vegas and Scottsdale. And now, as a result of that, the Colorado River dries up in the desert and never makes it to the sea." ∎

> **"Water belongs to everybody. It doesn't belong to Congress, the Senate, big corporations—it belongs to all of us. Everybody has the right to use it."**
>
> —ROBERT F. KENNEDY, JR., ATTORNEY AND PRESIDENT OF WATERKEEPER ALLIANCE

TOUR GUIDE

Can Technology Solve Our Water Problem?

If we're willing to pay the cost, it's already possible to turn ocean water into fresh water. One of the processes we use is reverse osmosis, which basically involves squeezing salt water at high pressure through a membrane that blocks the salt from getting through. Desalination, or removing the salt from salt water, can be accomplished on a large scale, and the resulting water can be piped to cities or farms far inland. Desalinated water costs about 60 cents per hundred gallons—that's twice as much as building a new reservoir or recycling wastewater.

DID YOU KNOW

Going into 2016, the entire state of California ran critically dry, with more than three-quarters of the state registering severe, extreme, or exceptional drought conditions.

"The United States leads the world in tornadoes, and we get these mondo hurricanes. We are [natural] disaster central. We've got everything but the frogs and the locusts, basically. We've got the fires, the floods—we've got all of that."

—DR. NEIL DEGRASSE TYSON

CHAPTER THREE

Where Do Storms Come From?

Tornadoes. Hurricanes. Floods. They're extreme weather features—air, water, and heat mixing in different proportions to produce powerful releases of energy that can wreak havoc on our lives.

When it comes to storms, heat drives the bus. As the sun shines, some parts of Earth (like the oceans) naturally absorb more light than others (like the polar ice caps). Temperature differences arise, and heat flows from warmer places to cooler ones. Storm fronts form where hot air and cold air collide. And if they happen to collide violently, right over your home, make sure your roof isn't leaking.

As with most things, storms aren't necessarily all bad. Hurricanes, for example, can bring lots of needed rainfall. Storms are just changes in the weather—until they get too severe. And the more heat there is to circulate and concentrate, the stronger a storm can get. We know Earth's surface is warming up. But by how much? Can—and should—we halt the heating? How much heat is too much? Hopefully, we won't have to find out the hard way.

Extreme-weather gear only goes so far.

What's the Difference Between Climate and Weather?

One classic way to understand climate is to think about the question of when you plant your tomatoes every spring. You plant by climate zone—an indicator that predicts the amount of rain they'll get in a given season, the typical date of the last frost, the average high summer temperature, the month when the leaves will change color, and so on. Those things are pretty predictable, year in and year out, based on climate.

Weather, on the other hand, is that late frost or heat wave or dry spell that forces you to take special care of your plants, or the sudden thunderstorm and lightning strike that knocks out electrical power to your house. These things change, hour to hour, day to day, and although the season when you're most likely to experience them might be regular, the day or night they happen is not.

We like to think that climate is predictable—but climate changes, too. Indeed, ironically, some people deny the effects of global warming by saying, "Why worry? The climate's always changing!" Well, not so fast—literally.

And it does matter. Since the dawn of agriculture, every society that experienced a significant climate change was either dispersed or destroyed, and died out. If climate change happens to us, will we share the same fate? ∎

> *"Weather is short-term, and climate is the average of weather over the longer term."*
> —ANDREW FREEDMAN, JOURNALIST

DRINK OF THE EVENING

Stormy Weather

Concocted by Neil and bartender Dirk at the Bell House

This concoction is a variation of the Dark 'n' Stormy, a classic Caribbean cocktail.

2 oz. rye whiskey
6 oz. ginger beer

Pour both ingredients into a tall glass, half filled with ice. Stir.

> *"The jet stream is the chaperone of weather at the junior high dance of climate science."*
> —EUGENE MIRMAN, COMEDIAN

THINK ON THIS ▶ Is the Weather Getting Worse?

When the climate changes, weather changes too—but it's only observable over long periods of time, and rarely with absolute certainty. By numerous measures, though, certain kinds of natural disasters have indeed become more common and more violent over the past several decades, especially those like droughts, floods, and hurricanes, which are driven by warmer air and water temperatures worldwide.

What Causes Severe Weather?

A lot of people seem to think the world is now experiencing more extreme weather events than ever. Astro-meteorologist Dr. Neil deGrasse Tyson wonders: "These intense storms . . . seem to be record-breaking in intensity and scale and size and damage . . . Where the hell do these storms come from?"

Let's start by talking about any storm, extreme or not. Weather happens through the interaction of warm air, cold air, and water. Pockets of warm air rise upward, leaving a zone of low pressure below it; other air flows into that zone to equalize the pressure: That creates wind. The flow isn't always smooth and gentle, though—and if the warm air is moist, the water vapor in it will condense as it cools. That creates clouds and rain. Static electricity builds up in the clouds as they fly past one another, discharging in bolts of lightning. And if the process feeds back on itself and intensifies, you get thunderstorms, microbursts, tornadoes, and hurricanes.

To feed their severity, storms need energy— and that usually comes in the form of heat. Just in its water vapor alone, a typical weather system contains heat energy equivalent to numerous atomic bombs! ■

TOUR GUIDE

What Were the Worst Hurricanes in U.S. History?

1900: Galveston, Texas—an island eight feet above sea level—was wiped out by a storm surge 15 feet high. About 6,000 people died.

2005: Katrina, a Category 5 hurricane, pounded the U.S. Gulf Coast. In its wake, levees protecting New Orleans, Louisiana, were breached, flooding the city. More than 1,800 people died.

2012: Sandy became the largest "superstorm" in Atlantic history, its diameter stretching out more than 1,100 miles. It killed 233 people in eight countries and caused $75 billion worth of damage.

THINK ON THIS ▶ How Does the Hurricane Intensity Scale Work?

Hurricanes are classified by their maximum sustained wind speed. Category 1, at least 74 miles an hour; Category 2, at least 96; Category 3, at least 111; Category 4, at least 131; and Category 5, at least 156. "I once looked up the description of the damage at each one of these intensity scales," says Dr. Neil deGrasse Tyson, astro-inferno-cist. "It was like the descent into hell in a Dante story line . . . This is Earth trying to kill us."

"Earth has never had a stable crust. Just look around; it's churning daily. That's why we have earthquakes. Go look at the U.S. Geological Survey earthquake page on the Internet. It's a record of all the earthquakes in the world. There are like hundreds a day. Every day."

—DR. NEIL DEGRASSE TYSON, ASTRO-QUAKE-ICIST

Certain coasts are at higher risk for tsunamis.

What Causes Earthquakes, Volcanoes, and Tsunamis?

Earth's crust isn't a single solid piece—it's made of huge tectonic plates of rock, some of them millions of square miles in size. The plates slide around on Earth's mantle at the rate of a few inches per year—a barely noticeable rate. But when the plates run into each other, they can get stuck, and pressure builds.

That's the starting point for all these huge events. First come earthquakes—but we never quite know when. "Earthquakes actually are sourced in the mantle, that fluid portion below the brittle crust of our planet," explains volcanologist Dr. James Webster. "To actually be able to predict at any given moment in time what's going to happen in the immediate future becomes difficult."

In the end, if one plate buckles or slips, or the pressure otherwise suddenly releases, the earth quakes. If the pressure releases violently enough, the liquid rock below the plates belches upward onto the surface, and a volcano erupts. And if an undersea quake disturbs the water just wrong, a wall of water can surge outward from the quake's location, causing a tsunami, or tidal wave, and sweeping away everything in its path.

Thankfully, we've developed technology that enables us to predict natural disasters and measure their intensity. But it took time to make accurate systems. "The old Richter scale could not measure earthquakes that were of the highest magnitudes accurately," says planetary scientist Dr. Steven Soter. "So a new magnitude scale, based on what's called moment magnitude, was devised in the 1970s . . . It turns out to be a factor of 33 [times more powerful] for each magnitude . . . in terms of the amount of energy release." ∎

Smog clouds daylight in Nizhny Novgorod, Russia.

How Does the Greenhouse Effect Work?

DID YOU KNOW

On Earth, levels of greenhouse gases like carbon dioxide change naturally over many thousands of years. Humans, though, have managed to double the carbon dioxide level in less than two centuries— a hundred times faster than it would happen in nature without humanity's help.

Lying in bed on a cold night, your body heat radiates upward, cooling you down. If you have a blanket on, its fibrous layers will absorb some of the heat, creating a barrier of warmth so that your body loses heat more slowly and stays warmer. If you have a whole stack of blankets on, you'll retain even more heat and get nice and toasty.

That's how the greenhouse effect works. It's a misconception that carbon dioxide, methane, water vapor, or other greenhouse gas molecules reflect heat like glass. Rather, they absorb and reradiate infrared radiation—which we feel as heat—really well. They don't absorb visible light nearly as well, so sunlight can shine down on Earth's surface with no trouble. But terrestrial heat from any source—be it warming from the sun, geothermal heat from Earth's interior, or humans burning fossil fuels—will leak upward and outward into space at a slower rate because of our blanket of greenhouse gases. ∎

What Effect Do Sunspots Have on Earth?

Sunspots are intense magnetic storms on the sun's surface, many of them much larger than our entire planet. Although the total electromagnetic energy output increases where the sunspots are located, the temperature in those spots actually drops slightly, so they look dark compared to the solar background light. They don't make much of an impact on the heat and light we get from the sun, but the resulting bursts of charged particles do make a difference here on Earth.

Dr. Judith Lean, a solar and atmospheric scientist, explains how these storms affect us: "The sun controls the weather of the environment where satellites orbit and where radio waves communicate . . . If you want to connect with your iPhone, and a big solar storm comes along, smashes all that plasma into the Earth's magnetosphere, and affects the satellite that is connecting your communications, then you do care about the weather in space."

In the end, sunspots don't usually last more than a week or two, and the area of the Sun's surface they cover is small, so their total effect on Earth is usually tiny. ■

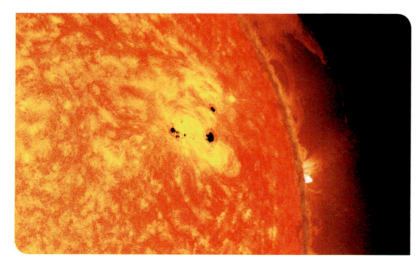

These black sunspots from 2013 measure more than six Earths across.

BACK TO BASICS

Are We Headed for Another Maunder Minimum?

The frequency of sunspots ebbs and flows, but every 11 years or so, the number reaches a peak—astronomers call that solar max. Then it dips down, reaches a minimum, and rises back up again. The most recent solar max was a dud, with barely half the number of spots compared to previous times. What's up with that? "This may be one of those unusually long cycles, or we may be going into something like the Maunder minimum, back in 1645, when sunspots disappeared for about 50 years," says Dr. Stephen Keil, solar scientist.

At the time of the Maunder minimum (named for two scientists, Annie and Walter Maunder, who first explained the period), temperatures worldwide also dropped slightly, the winters were longer than usual, and some historically ice-free rivers and lakes froze over. But that's not happening now—we're getting warmer.

DID YOU KNOW

Less than one percent of the Sun's surface is covered with sunspots. Other stars are known to have a surface more than 50 percent covered with spots.

Polar guide Eric Philips trekking in Siberia

Cosmic Queries: Planet Earth

Is It Getting Cold Around Here?

"There are some places that are not as susceptible to natural disaster . . . What you want is to be in a place that is not as susceptible to the fluctuations in climate. And among those places would be rain forests, for example."

—DR. NEIL DEGRASSE TYSON,
ASTRO-JUNGLE-IST

Serbian astrophysicist Milutin Milankovitch (1879–1958) made a number of detailed calculations that showed that tiny long-term variations in Earth's motion as it orbits the sun can affect our climate in subtle yet important ways. The Milankovitch cycles have three main components: First, Earth precesses—that is, it wobbles like a top that's starting to fall—so the direction the North Pole points keeps changing.

The Earth makes a full wobble once about every 23,000 years. Then there's the tilt angle of Earth's axis, which affects the lengths of the seasons and changes by about two degrees every 40,000 years. Third, the shape of Earth's orbit around the sun changes, which affects the amount of sunlight Earth receives over the course of each year. It can change from slightly elliptical to nearly circular and back again about every 100,000 years.

Right now, Earth seems to be in the middle of the range of these cycles—so by Milankovitch's reckoning, our climate should be relatively moderate. ◼

Climate Cycles: Where's the Proof?

Earth's climate changes naturally, both locally and globally.
These changes happen over long periods of time, though,
whereas human-produced climate change has been rapid. What kinds
of evidence can we use to measure these changes and get the big picture?

|||||||
◀ OCEAN SEDIMENT

Fossilized plankton, diatoms, and other organisms buried deep beneath the ocean can show the environmental temperatures and conditions during which their bodies were built.

|||||||
▶ POLLEN

When these tiny grains are washed or blown into lakes and ponds, they get embedded in layers of sediment and provide a record of long-ago plant life—and the climate at the time that would support them.

|||||||
◀ ICE CORES

Tiny bubbles of air trapped in old, deep glaciers and packed ice preserve the greenhouse gas levels in Earth's atmosphere going back more than 500,000 years. We can track changes in those levels from then until now.

||||||
◀ ROCK

Wind-driven soil and sand deposits, such as loess and eolian dust, form during ice ages around the edges of massive glaciers. Layer by layer, they tell the history of some of Earth's most dramatic climate changes.

||||||
▲ CAVES

Mineral deposits formed from groundwater in underground caverns, like stalagmites and stalactites, record the climate in which they formed through the properties of their atoms and the thickness of their deposits.

||||||
▲ LAKE LEVELS

In drier parts of the world, the depths and areas of lakes can change dramatically when climate and moisture levels change. Fossil deposits along their shores trace that climate history.

||||||
▶ TREE RINGS

Paleoclimatologists can measure the environmental conditions and fire history of different parts of the world from the growth rings and fire scars of trees, as well as from charcoal in sediment records.

StarTalk Live!: Climate Change

What Happens If Earth Warms by One Degree Celsius?

Based on climate-modeling research, scientists think Earth's average surface temperature will go up somewhere between 1°F and 5°F over the next 50 years—most likely somewhere in the middle, at 1°C (1.8°F).

Two big unanswered questions are: What changes will that heat produce, and where will those changes occur? Scientists and policymakers on the United Nations Intergovernmental Panel on Climate Change (IPCC) are trying to figure that out, using decades of data and research.

Their discussions are anything but quiet, says climatologist Dr. Cynthia Rosenzweig: "IPCC chapters have very spirited debates before they come to consensus."

And while comedian Michael Che wonders if anyone's been shot at these debates, the simple answer is: No, thankfully.

"I'm just saying," Che argues, "these disagreements can get out of control if we're not careful."

On at least one scientific fact, everyone totally agrees. One degree Celsius won't feel like much on any given day to any given person, but totaled across the globe, it's enough extra heat energy to power hundreds more hurricanes, floods, tornadoes, and blizzards every year. One thing leads to another, and some such cycles speed up climate change, a process called positive feedback. ■

BACK TO BASICS

What's the "Hockey Stick"?

Imagine a graph that starts rising slowly and then shoots upward: That's the "hockey stick." One dramatic version graphs the carbon dioxide level in Earth's atmosphere over time. For at least 400,000 years, the CO_2 level cycled between about 150 and 300 parts per million. Then in less than 0.02 percent of that time—just 70 years—the level shot up to 400 parts per million. The shape of the graph goes from a mostly horizontal line to a nearly vertical line almost instantaneously.

"Since 1880, there's been almost a degree Celsius warming . . . Up at the poles it's much, much more than that . . . on the order of three or four degrees Celsius . . . That's why you're seeing that melting . . . You get a positive feedback of warming."

—DR. CYNTHIA ROSENZWEIG, CLIMATOLOGIST

THINK ON THIS ▶ What Is Albedo, and Why Should We Care?

Albedo (rhymes with "torpedo") is a term for how much heat and light a surface absorbs or reflects. A perfectly black surface has albedo zero; a perfectly reflective mirror has albedo one. Fresh snow's albedo is about 0.8; ice's about 0.4. Soil and water absorb light and heat, so they rank about 0.1. If Earth's snow and ice melt, ground and ocean surfaces increase, the combined albedo drops, and Earth warms up faster.

"The mayors of cities around the world . . . have been getting together, and forming networks and signing on pacts with each other to set targets and timetables for greenhouse gas emissions."

—DR. CYNTHIA ROSENZWEIG, CLIMATOLOGIST

Melting global ice sheets will flood Manhattan.

Climate Confusion

How Can Climate Change Cause Both Droughts and Floods?

Extra heat is a versatile thing. Some hot places on Earth are deserts; others, tropical rain forests. When a hot place gets hotter, drought often results. And when a cold place heats up, its ice melts.

▶ HOW DOES MOISTURE RAMP UP CLIMATE CHANGE?

With ice melting and releasing liquid water, big climate change can occur. "When climate warms, even more moisture shows up, and then . . . it can rain heavier," says climatologist Dr. David Rind: "We've been seeing that for the last decade . . . heavier episodes of precipitation, more droughts, more floods, just an amplification."

▶ ARE THE SEAS ALREADY RISING?

You can ask the residents of South Florida, where ocean fish are washing up regularly on the streets of downtown Miami. Here's another data point, from climatologist Dr. Cynthia Rosenzweig: "These are the knock-on effects of warmer ocean temperatures . . . We have sea level rise, again, already happening. Just in New York City, we've had over the past 100 years, over a foot of sea level rise."

▶ DID CLIMATE CHANGE CAUSE SUPERSTORM SANDY?

No one storm can be linked directly to climate change, but the nature of a single storm can. "When any storm . . . comes along, the seas are [already] one foot higher, [so] the flooding goes one foot further in elevation," explains Dr. Rosenzweig. "So that extent of flooding that we had in Sandy, that is the actual direct link to climate change." ■

> *"The goal here really ought to be . . . a future where there's actually limitless energy, and we wouldn't even have to have these conversations, and we wouldn't have to be warming up the atmosphere, and having Earth do things that it hasn't done in a million years."*
> —DR. NEIL DEGRASSE TYSON, ASTRO-OPTIMIST

StarTalk Live!: Climate Change

Have We Reached the Tipping Point Yet?

Some research suggests there is a tipping point: a point at which, once Earth gets to a critical level of greenhouse gases or loses a certain amount of its polar ice, we will never be able to halt or reverse global warming. Is this possible? If so, are we there already?

Predicting the future accurately is always tough, even if you use the best tools available. And good scientists, like climatologist Cynthia Rosenzweig, readily acknowledge their limitations. "We use models, which are big mathematical sets of equations for the whole climate system . . . Are they doing a good job projecting? In general, they're doing a pretty good job, but some of these changes have been occurring faster than have been projected . . . Probably one of the most important examples is the melting of the polar ice caps."

Most climate change models in use today suggest that if a theoretical tipping point exists, we may be close to it, but we're not there yet. None of the models, though, can predict what scientific advances—or crazy natural events—will happen that could radically alter those predictions. ◾

Human activities are pushing Earth to the edge.

TOUR GUIDE

How Is Agriculture Affecting Climate Change?

We tend to point to big industrial smokestacks and city traffic pollution as the causes for global warming and climate change, but what about the way we grow our food? "[Agriculture] itself is an emitter of greenhouse gases," says Cynthia Rosenzweig, a climatologist. "The livestock also, because of the enteric fermentation [the way cattle digest their food] puts a lot of methane out, which is a very powerful greenhouse gas . . . Creating nitrogen fertilizer is highly energy-intensive. Plus the fertilizer itself, when it's applied, emits another greenhouse gas: nitrous oxide."

So making food is contributing to global warming. And climate change is changing how we make food. Droughts are destroying the productivity of agricultural lands. Once-fertile coastal regions are now underwater or uninhabitable. And planting seasons are changing, affecting which crops can be grown. The list goes on.

StarTalk Live!: Climate Change

Fossil Fuels and Greenhouse Gases: What's Really Going On?

There are plenty of intelligent and well-educated people who just can't believe human activity causes climate change. Take the experience of this hapless (name withheld for privacy) individual: "A few years ago, I'm at the doctor's office, and he's sticking a you-know-what slowly into my you-know-where, and he says, 'I know you're an astrophysicist, and I have to tell you, I just don't believe that people can have an impact on the climate, because we're so small and Earth's so big. What do you think?' "

So what's the right way to answer this deeply probing question?

Car fumes are just one contributor to climate change.

▶ LOOK AT EARTH

Look at pictures of Earth from space at night. See how we puny humans have turned darkness into daylight all over the globe? Why wouldn't we be able to add a little heat, too?

▶ DO THE MATH

The amount of carbon dioxide burning fossil fuels produces is easily calculated, and the numbers add up to how much CO_2 has been added to the atmosphere over the past century.

▶ LOOK AT VENUS

Do what astronomers do—check out our neighbor, Venus. The average temperature on Venus is 900°F. Its storm winds blow up to 400 miles an hour. There's no liquid water and no life on Venus. That's the real impact of a greenhouse gas. ■

"So what we're doing with the greenhouse gas emissions is like we're putting on a thicker, fluffier blanket. Sometimes we describe it that way. And that gets to impacts. That's why we care. Because the climate system affects every single thing on the planet."

—DR. CYNTHIA ROSENZWEIG, CLIMATOLOGIST

THINK ON THIS ▶ Could Cleaning the Air Increase Global Warming?

"In addition to putting all this carbon dioxide in the atmosphere . . . [we've also] put a lot of dust and dirt and aerosols in the air—and that reflects sunlight back to space, and it cools the climate. So the problem is, we don't know how much warming we've been doing relative to the cooling that these aerosols do."
—Dr. David Rind, climatologist

"I think the general attitude should be that we shouldn't waste the bounty of the Earth; that's the main thing."

—SIR DAVID ATTENBOROUGH, NATURALIST

CHAPTER FOUR

Is There a Solution for Pollution?

Plants and animals live, grow, and produce waste. In that sense, human pollution seems perfectly … natural. But just as industrialization has brought tremendous benefits to our species, the impacts of its waste extend far beyond mere inconvenience. Now our very existence seems to be threatened by the environmental changes we have wrought!

Happily, we are learning to clean up after ourselves. But as soon as we solve one dilemma, we have to tame a new monster. Until petroleum was first mined in the mid-1800s, we harvested oil from whales and they were being hunted nearly to extinction. Today we harvest fossil fuels from underground—and we have earthquakes caused by fracking. New York City streets were awash in horse manure until the advent of the automobile—and now we have smog and global warming.

What's our next step toward a healthy, pollution-free world? Is another solution coming? And what problems will come with it?

It is time to switch to renewable energy sources.

What's the Biggest Engineering Challenge Facing Us Today?

The first great engineering challenges were probably those involved with building the great monuments of ancient civilizations, like the pyramids of Egypt. Huge public works—like the Roman aqueducts or, in modern times, Hoover Dam—came next. All of these endeavors took great minds—and great resources—to think up and execute their completion. And our current engineering challenges are equally substantial at every scale, from nanotechnology (like using cell-size robots to cure disease) to biotechnology (like creating artificial organs and limbs), all the way to human habitation in space (like the International Space Station—and maybe, one day, Mars). But in terms of sheer size, what has been the biggest challenge?

Today, traveling away from our planet and into space—then making it back to Earth safely again—may be our absolute greatest engineering challenge yet. And hundreds of scientists and entrepreneurs are working to reach this major milestone today.

> "I'd say it's climate change. So how are you going to [fix] that? I would say it's going to be engineering the whole Earth. Thinking of the whole Earth as a system and getting people to work together to manage that system."
> —BILL NYE THE SCIENCE GUY

An engineer's job is to solve problems using both science and technology. So looking forward, what's humanity's biggest problem right here on Earth?

If it's global environmental degradation, including global warming-induced climate change, then the solution has to be created globally, too. The whole idea of climate change is too big to swallow in one gulp—and requires working together and thinking ahead. Like all good solutions to complex problems, everything needs to be divided into small, manageable pieces, each of which can be solved only while being informed of all the other related issues. Otherwise, the solution may have unintended consequences and cause new and even bigger problems than before. ∎

Global warming won't be solved with a cover-up.

> "When you talk about some of these solutions—like let's geoengineer something, let's design some organism that sucks CO_2 out of the atmosphere—what if we're too successful and we can't turn it off, and all of the CO_2 goes away? We would die . . . You need a certain range of CO_2, and you don't want to go too far out of that range in either direction."
> —DR. DAVID GRINSPOON, ASTROBIOLOGIST

StarTalk Live!: Storms of Our Century (Part 2)

Where Do Fossil Fuels Come From?

P resent-day reserves of coal, oil, and natural gas were formed deep underground from living organisms that died long ago. Their carbon content was slowly changed into chemical forms that release lots of heat—and carbon dioxide—when they're burned. Let Neil explain:

"During the Carboniferous period in the history of Earth . . . trees would die and they would stay there forever . . . Vegetation is made of carbon—that is the principal ingredient—so every tree that grew took carbon out of the atmosphere. [When the tree fell,] the carbon stayed with the tree. This continued, and continued, and huge layers of dead vegetation got submerged in the crust of the Earth. That vegetation became fossil fuels. And so the balance of carbon in our atmosphere is disrupted when we take carbon from a place where it has been buried for millions of years and introduce it into the stable balance of carbon in our atmosphere today." ■

DID YOU KNOW
The seven largest emitters of fossil fuels—the United States, the European Union, China, Russia, Japan, India, and China—contributed more than 70 percent of energy-related CO_2 emissions in 2004.

TOUR GUIDE

Can Airplanes Fly on Alcohol?

Neil explains Brazil's unique approach to cutting back on fossil fuels: "Brazil has the third largest aerospace industry in the world. It's a $20 billion industry that employs 18,000 people. They invented an airplane that runs on alcohol, pure alcohol . . . which is essentially solar-powered. Because alcohol is derived from plant products, and plant products get their energy from the Sun."

But there are issues. Alcohol is produced from plants—in Brazil, mostly sugarcane—but processing the plant material into alcohol and then into fuel actually requires lots of energy, and the result doesn't provide as much energy as jet fuel. So for now it doesn't end up a saving proposition, but further work on the technology could make alcohol-powered planes feasible worldwide.

"So here we are drinking alcohol on our airplanes, and they're making airplanes that can run on alcohol."

—DR. NEIL DEGRASSE TYSON

Were Prehistoric Sunsets Red?

Some people say that our red sunsets are caused by man-made particulates—that is, air pollution—scattering the light, more now than ever. So that raises the question: Would pre-pollution sunsets have been really boring? Not at all. To begin with, why does the sky look blue? It has to do with the atmosphere. If Earth didn't have an atmosphere, the sky would be black, even in the daytime. But we do have an atmosphere, full of air molecules, and as sunlight penetrates our atmosphere, those molecules

"It's not only pollution. It's pollen, it's water vapor, it's dust kicked up from deserts, it's volcanic particles—all those particles make a sunset red."

—DR. NEIL DEGRASSE TYSON, ASTRO-SUNSET-IST

scatter light in all directions. Turns out they scatter blue light much more effectively than red light, so it's mostly blue light that hits our eyes. As the Sun sets, its light is traveling through more air molecules to meet your eyes, and also past suspended particles that are larger than air molecules. The scattering of sunlight through the combined extra air and particulate matter make the sky at sunset look yellow, orange, and red. Any source of particles would do the trick. ∎

An illustration of Utahraptors running across a sunset-lit prehistoric grassland

StarTalk Live!: Water World (Parts 1 and 2)

What Are Energy Incumbents and "Biostitutes"?

ncumbent energy sources are a lot like incumbent politicians: They are already in place, we know they can get the job done, and sometimes people want it to stay that way. Energy incumbents include coal, oil, gas, and nuclear—and many people count on them sticking around for financial reasons: jobs and profits. But do we want to vote them out and vote in new energy sources such as wind, solar, and geothermal?

> *"In a perfect democracy, you would have policy that was driven by science and by objective criteria and by empirical data."*
>
> —ROBERT F. KENNEDY, JR.,
> ATTORNEY AND PRESIDENT OF
> WATERKEEPER ALLIANCE

"One of the things that environmentalists and people in the business community have come to realize is that good environmental policy is identical to good economic policy," says attorney Robert F. Kennedy, Jr., president of Waterkeeper Alliance. "If you talk to the big polluters, they'll tell you that we have to choose between economic prosperity on one hand and environmental protection on the other, and that's a false choice."

For those who seek to obscure scientific reality to further the aims of a corporation, Kennedy has coined a derisive name—a mix of "biologist" and "prostitute." "By investing a little bit of money in a politician, and then some money in these so-called 'free market' think tanks . . . they'd fill them up with these phony scientists we'd call 'biostitutes,' who will say there's no such thing as global warming." ■

TOUR GUIDE

What Were They Thinking?

"I got a film the other day of an albatross, a nesting albatross that had just circled the Antarctic Ocean and the Antarctic continent, collecting food for its young . . . Everything, everything that it put in that chick's mouth was plastic. That happens everywhere, all over the Pacific. Birds that are collecting food from the surface are taking that back and giving [it] to the chicks. And it's plastic, and it will be there forever. We said, 'It's wonderful; we've invented a new compound that is indestructible'. . . And nobody said, 'Oy, if we are going to keep on doing that, what's going to happen to that?' Isn't that extraordinary?"
—Sir David Attenborough, naturalist

DID YOU KNOW

Plastics are estimated to make up almost 80 percent of marine debris in the world's oceans.

THINK ON THIS ▶ Could Magma Clean up the Pacific Garbage?

"That's the kind of creative thinking that we really need. How do you really get rid of plastic? Sure, put it in a subduction zone and send it to the mantle of the Earth. I'm all for it, if I could see the details . . . There's nothing that sounds wacky about it to me."
—Dr. David Grinspoon, astrobiologist

Can't We Just Plant More Trees?

Trees (and other plants) take in carbon dioxide and release oxygen. So if too much CO_2 is warming the globe, wouldn't more trees help? The Buddhist spiritual leader His Holiness the Gyalwang Drukpa, began a remarkable project a few years ago: Annually, thousands of volunteers gather with young saplings and plant trees in the Ladakh region of the Himalaya. In 2012 alone, they planted nearly 100,000 trees in a single hour!

The act of planting trees is simple, and when done wisely in a coordinated matter it can change an ecosystem for the better. Like any single action, of course, the effects of tree planting are limited, both ecologically and in time; the trees will take a long time to grow by human standards. And planting new trees may also have unintended consequences: Will they crowd out existing plant life by outcompeting them for resources? Would a large area covered with a single type of plant lead to increased disease and pest damage? Ultimately, though, events like these are a great way to start a good thing. ■

More trees won't be the only answer.

BACK TO BASICS

Bill Nye's Words of Wisdom About Climate Change

"It's going to be a close call. It's not clear who is going to get to the buoy first: the climate change that undoes so many of our urban populations and causes enormous upheaval, or the scientifically literate populace that is emerging with an enthusiasm for scientific truth, and understanding the world by critical thinking . . . And so when I say it's a close call, I mean: We could lose. Humankind could really be in huge trouble. Not just the coastal populations in the developing world, but everybody could be in trouble. So this is why . . . it's a race. This is why we do what we do."

—Bill Nye the Close Call Guy

"When you pump large amounts of water underground, you loosen up some of those faults, and they move, and you've got earthquakes. Fortunately, the earthquakes you get from [fracking] tend to be shallow, and rather small. And there's even the argument that that's good, because they're relieving tension on the faults, which otherwise would build up and eventually result in a larger earthquake . . . It does speak to the fact that there are unintended consequences when we start . . . altering the Earth in intense ways."

—DR. DAVID GRINSPOON, ASTROBIOLOGIST, ON IF FRACKING CAUSES EARTHQUAKES

The Story of Life on Earth With Sir David Attenborough

Can We Tax Our Way Out of Pollution?

I n much of the world, carbon credits are a way of life. A government or business can get financial credit for steps taken to reduce CO_2 emissions and can use that credit to offset other carbon-producing processes they choose to continue. It's a way to keep CO_2 emissions at a lower level. Carbon credits have become an integral part of the economic and ecological dynamic in some countries. Could they come to the United States?

Bill Nye starts us down this road: "Oil companies, fossil fuel companies, already have this built in. They're anticipating a $40 per ton [of carbon dioxide emitted] surcharge that's in their economic modeling . . . If we did this, and then, the countries that don't produce as much carbon dioxide would have a tariff or a fee on the goods that were being imported that had a high carbon. This is not 'everything is going to be fine'; it is part of the solution." ∎

DID YOU KNOW
Cooperation to reduce climate change among the major carbon-producing nations—China, the United States, the European Union, India, Russia, Indonesia, Brazil, Japan, Canada, and Mexico—could cover more than 70 percent of greenhouse gas emissions around the world.

EXPERIENCE THE GRAVITY OF
HD 40307g A SUPER EARTH

BACK TO BASICS

If We Screw Earth Up, Can We Go to Another Planet?

As Neil tweeted, "Mysteries of #Interstellar: Can't imagine a future where escaping Earth via a wormhole is a better plan than just fixing Earth." From a purely scientific standpoint, it would take much, much less energy to repair our world than for all of us to travel to another one. Bill Nye chimes in: "This tradition of just, you've trashed an area, just move to a different area—just expand, just march on . . . We're going to have to be better stewards of the environment and, dare I say it, Neil, change the world."

THINK ON THIS ▶ **Can GMOs Save Our Environment?**

"The place to start is making . . . bacteria that could metabolize oil in an oil spill. That would be a really cool application . . . You could have some bacteria that could metabolize [the organic chemicals], and then leave behind all of the other stuff that comes out of oil wells—sulfur and so on—and let it sink to the bottom of the ocean." —Bill Nye the Science Guy

Is the Climate-Change Debate Fair and Balanced?

Journalists believe in giving both sides of a story. In science, rival hypotheses are considered, too—until a clear consensus emerges due to overwhelming evidence. Science correspondent Miles O'Brien describes where we are on the topic of climate change: "Some stories have one side that is represented by, say, 95 percent of the scientific community in the world. Is it fair, in a story about climate change, which I'm obviously talking about, to do this classic journalistic convention of 'equal time for both sides'? . . . Is that serving the truth? I would submit to you, not. As a matter of fact, that is feeding obfuscation; that is actually perpetuating a myth—dare I say, a lie . . . The scientific jury is in, here."

On the subject of climate change, the data clearly show that Earth is warming up, and that humans are contributing to the warming. That's the fairest assessment of the evidence. ■

"There is no more scientific debate. There's a political debate; there's a debate over money —over how we should spend it, what we should do. But there is no scientific debate. Let's just get over that."

—MILES O'BRIEN, PBS SCIENCE CORRESPONDENT

BIOGRAPHY

Why Should the Pope Address Climate Change?

Pope Francis, born Jorge Mario Bergoglio in Buenos Aires, Argentina, worked as a chemical technician before joining the priesthood. When the Pope speaks of climate change and its harmful effects on the poor people of the world, some are skeptical of a religious leader interested in science. Others applaud him. "Reason and faith are not inconsistent. Logic and faith are not inconsistent," says Rev. James Martin of the Society of Jesus. "Pope Francis, who starts with faith—he's naturally interested in the natural world. He wants to know more about God's creation, and so why not study chemistry?"

THINK ON THIS ▸ **Do Disaster Movies Move the Debate?**

Maybe a movie depicting a bleak future caused by global warming will sway people. "These sorts of events lend themselves to what people call, in academia and elsewhere, disaster porn," explains journalist Andrew Freedman. "It's easier to go for the big stuff than to be subtle." The challenge is to get people interested while staying truthful. Every exaggeration makes real threats that much less believable.

Is the Solution to Stop Burning Fossil Fuels?

Some say we should banish fossil fuels from our future. "[They say] we have come to demonize the consumption of energy . . . That's the wrong attitude, in my opinion," says astro-energy-ist Dr. Neil deGrasse Tyson. "If you're going to demonize something, demonize that which alters your environment. The universe has limitless energy . . . I'd be embarrassed to tell an alien, who just moved through the vacuum of space bathed in limitless starlight, that here on Earth we kill one another to extract oil . . . Is the solution to stop burning fossil fuels, or is there some other solution that comes along, that doesn't negate this, it just renders it completely obsolete, as the car did to the horse?"

▶ WHAT GOOD CAN COME FROM BURNING OIL?

Actually, a whole lot of good comes from burning oil. Among other things, oil keeps us warm in the winter and cool in the summer, it gets us to places near and far through various means of transportation, and it allows us to be productive and contribute to the local and global economy. The problem is in the medium term and long term—a couple of human generations or more from now—when the environmental consequences of burning that oil threaten to undo all of that short-term benefit.

▶ WHY DON'T WE JUST INNOVATE OUR WAY OUT OF USING OIL?

Plenty of people are working toward that very goal. Including entrepreneur Elon Musk, founder of Tesla Motors and SpaceX: "If we don't find a solution to burning oil for transport and we then run out of oil, the economy will collapse and civilization will come to an end as we know it . . . We have to ultimately get off oil no matter what." ∎

"Why would you run this crazy experiment of changing the chemical composition of the atmosphere and oceans by adding enormous amounts of carbon dioxide that have been buried since the Precambrian era? That's crazy. That is the dumbest experiment in history by far."

—ELON MUSK,
FOUNDER OF TESLA MOTORS AND SPACEX

Did the Big Bang Make Technology Inevitable?

"At the big bang, you have all of this energy in a very small volume, and it is so hot, it is so energetic that matter is forming out of energy . . . So what we have here is this soup, it's a matter-anti-matter energy soup."

—DR. NEIL DEGRASSE TYSON

I n the 20th century, some scientists began to contemplate the anthropic principle—an idea that our universe is the way it is because if it weren't, we wouldn't be here. Does that mean that the birth of our universe *requires* that we do what we've done? Is it possible that everything that has happened since the beginning of our universe—since the big bang, essentially—has been ordained?

That would mean that every invention and discovery, from man controlling fire to the first wheel and then the first automobile to typewriters and then lightning-fast computers, was inevitable from the beginning of time. Many bright researchers, including astrobiologist Dr. David Grinspoon, have dared to dive into this hard-hitting question. Here's his best answer:

"The laws of nature were set into motion by the big bang, and those laws seem to be conducive to the evolution of life on some planets. And I think on some of those planets, that complex life will lead to technology. So in a certain sense, technology was ordained by the big bang. Not exactly on this planet in exactly the way it is . . .

"And yes, I think technology can be used to heal the problems we have. Not technology alone; a lot of it is going to come through self-knowledge and our being able to manage ourselves more wisely. But that knowledge of self has to go hand in hand with knowledge of nature and knowledge of how to manipulate nature, which is technology. So technology, yes, it was ordained in the big bang, and, yes, it will be part of the solution. So thank you, big bang." ■

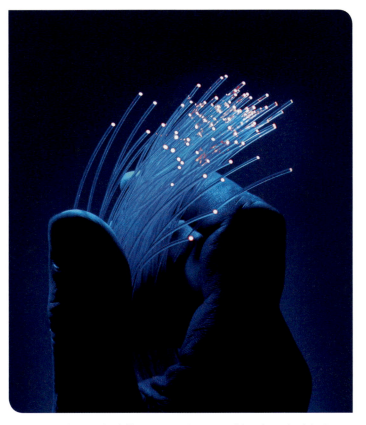

Maybe optical fibers were inescapable after the big bang.

"Think about how a forest responds to the seasons . . . The whole year is a series of changes . . . If we look at a city, well, how does your city really change to the seasons? . . . We've tended to accept that our cities are as they are. They don't have to be this way."

—MELISSA STERRY, FUTURIST

Can We Afford Electric Cars and Solar Panels?

Going all-electric would be great. It costs a lot to retool for the future, though. Dr. Neil deGrasse Tyson, astro-budget-ist, asks: "So, do you have a plan? What's your plan? Because as long as oil is cheap and it's cheaper than my solar panels, how do you expect people to [afford it] . . . If you're rich, you can buy the car that saves gas, that costs you more than the car that doesn't save gas."

Neil posed the problem to his friend Bill Nye the Science Guy: How can we possibly afford electric cars and solar panels?

Bill Nye replied: "To maintain our current way of life, can we afford *not* to have them?"

People complain about emission standards and other such rules. "The big thing as we say about climate change, is: If you are opposed to government regulation now, you don't like governments now, just wait till stuff gets bad," Nye says. "Just wait till Floridians have to abandon their homes and Miami is half underwater—and then there's going to be regulation." ■

BACK TO BASICS

How Will We Solve Energy Storage?

To compete with fossil fuels, solar- and wind-power generation needs to be consistent, reliable, and controllable. But skies get cloudy and winds stop blowing—and not always when you expect it. And electric cars can only go so far, because they can't store enough energy between charges.

Both of these problems can be solved with rechargeable batteries that are small and light, and hold a lot of charge. But we're definitely not there yet. Current electric vehicles have a range of about one mile for every four to eight pounds of batteries in the car, and big power stations have to be able to hold and distribute much more power at any one time than a warehouse full of such batteries could manage.

Optimistic projections suggest that at least one new battery technology, the lithium-air battery, could replace the current lithium-ion battery in about 10 years. That would result in an 80 percent weight reduction. A lot can happen in 10 years, though.

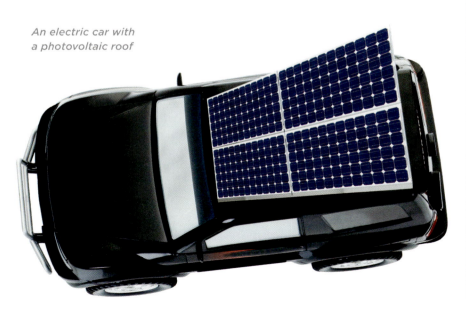

An electric car with a photovoltaic roof

"You read about the Mayan calendar, you've heard about the Apocalypse, and your brain just connects the dots—and then you become certain that the world will end."

—DR. NEIL DEGRASSE TYSON, APOCALYPS-ICIST

CHAPTER FIVE

Is This the End of the World as We Know It?

Armageddon. Ragnarok. The "end of days." Why, in the annals of superstition, myth, and prophecy, is humanity's collective demise such a popular theme? Psychologists who study apocalyptic impulses say that if we've all gotta go sometime, most of us would rather all go down together, in a blaze of glory, than alone, unnoticed, and forgotten.

In early cultures, it was natural disasters that destroyed everything. As we started understanding the forces of nature, people turned to the supernatural, like the devil, who was supposed to be let loose upon the world in the year 1000. Today, religious motifs are being replaced by pseudoscience—people mixing parts of scientific concepts and plausible-sounding statements to create fictional scenarios that only slightly resemble reality.

Yes, we face real dangers, for which we can study and prepare. How can we tell the difference between fact and flimflam? Armed with scientific literacy, we can look to the future unafraid.

Are we looking at an apocalypse
on planet Earth?

The End Is Nigh–or Is It?

Some people anticipate the return of the biblical Jesus as a joyous event marking the end of the world. But Neil certainly doesn't: "This exercise in judging when Jesus is going to come back is one of abject failure in the past, so to continue this going forward would be futile in my judgment."

The return of Jesus is only one of many renditions about the end of the world as we know it. Here's a list of other recent—and, yes, wrong—apocalyptic predictions.

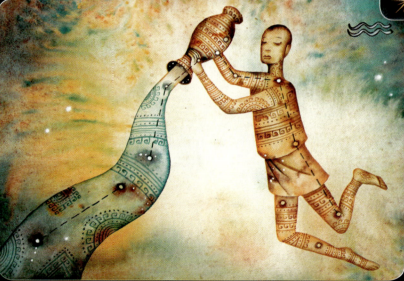

‖‖‖‖

◄ 1962 PLANETARY ALIGNMENT

A planetary alignment coincided with an Earth-Moon perigee, and high winds created storms and high tides in the eastern United States. Despite disclaimers from meteorologists, many people extrapolated flood warnings to mean global doom.

‖‖‖‖

▶ 1997 HEAVEN'S GATE

The beautiful passage of comet Hale-Bopp led to the tragic group suicide of 39 members of the Heaven's Gate cult, who believed they would be transported to a spaceship hiding in the comet's tail.

◄ 2000 Y2K

Computers unable to handle the year change from 1999 to 2000 were expected to fail and predicted to lead to catastrophes everywhere from banks to airports to nuclear power plants. Millions of dollars were spent on "Y2K" upgrades. Ultimately, 1/1/00 came and went.

Comet Elenin

▲ 2011 COMET ELENIN

As comet Elenin approached, some claimed it was actually a rogue planet named Nibiru, whose arrival would cause planetary destruction. The comet broke into pieces before ever reaching Earth.

▶ 2012 MAYAN CALENDAR

An ancient Mayan calendar fragment was said to tell of the end to a "cosmic cycle" on December 21, 2012. A wide variety of ways in which the world would end on that date circulated. None of them happened.

▶ 2015 BLOOD MOON PROPHECY

Two Christian ministers claimed that the "blood moon prophecy" from the Book of Revelation, signaling the end-time, would come to pass during the lunar eclipse of September 27-28, 2015. It didn't.

What Would Abe Lincoln Do?

There's an interesting story about Abe Lincoln and his response to a Leonid meteor storm that occurred in 1833. Neil tells the tale: "The local preacher was well-read in the Bible, of course, where in the Book of Revelation it says at the end of days, stars will fall from the sky and land on Earth. So he sees this, he comes running, knocking on doors, goes into Abe's room, and says, 'The end of the world is near. Repent.' Lincoln comes running out. He looks up, sees this beautiful meteor shower, but also notices that all the stars that he's familiar with are still there. The Big Dipper. Orion. So he goes back in and goes back to sleep . . . It helps to know a little astrophysics. He was well read—better read, apparently, than the preacher." ■

Can We Survive a Magnetic Polar Reversal?

Magnetic fields, like magnets, have north and south poles. Over long periods of time, magnetic fields can switch their polarity—north becomes south, south becomes north. The sun does this once every 11 years—and based on the fossil record, Earth appears to have done it hundreds of times in the past, most recently about 800,000 years ago.

Recently, after scientific satellites measured Earth's magnetic field to have changed slightly, doomsayers began to claim that a magnetic polar reversal was on its way—and at the moment of the flip, we'd lose our magnetic protection from solar wind, and life on Earth would be wiped out.

Is this likely? No. Just because the poles reverse doesn't mean the magnetic field goes away—it just changes orientation, and it happens gradually and erratically. Even if a short dip or switch in field strength were to occur for any reason, we'd still have enough protection from Earth's atmosphere to weather the solar wind. It's just a natural phenomenon—a neat feature of our planet, not a problem. ∎

> "You can look at the fossil record. Last time around, while we had zero magnetic field, they carried on making babies just fine. So it does not appear to be as severe as you might think, the evidence shows."
>
> —DR. NEIL DEGRASSE TYSON

Earth's magnetic field is more complex than a horseshoe magnet.

Will There Be an Anti–Big Bang?

Decades ago, we knew the universe was expanding, but we didn't know if it would expand forever or stop and reverse in a "big crunch," a catastrophic collapse of the cosmos. The confluence of three parameters would tell the tale: (1) the current expansion rate, (2) the density of the universe's matter, and (3) the cosmological constant—a theoretical repulsive effect originally proposed by Einstein.

Beginning in the 1990s, astronomers using telescopes like Hubble, WMAP (the Wilkinson Microwave Anisotropy Probe), and Planck were able to measure these three parameters with increasing accuracy, so today most agree that a big crunch will not occur.

"All data show that the universe will continue to expand forever and not slow down and recollapse, which is philosophically upsetting to many," says Dr. Neil deGrasse Tyson, astro-let-down-icist.

Upsetting? Maybe, if you're a fan of the universe having an expiration date. On the other hand, if you like the idea that time will go on forever, the scientific reality may actually be a comfort. ■

Many have theorized premature human expiration.

BACK TO BASICS

What's Joe Rogan's Big Bang Machine?

"I had a bit about a 'big bang machine.' The idea was that scientists have never figured out what started the big bang. And then I think that 14 billion years ago there were some scientists . . . and one day, they made a big bang machine. One guy sat around and said, 'I'll press it'—and he hit the button, and the whole thing restarted. And that is the cycle of humanity: It goes from single-celled organism, to multicelled organism, to conscious entity, to autistic dude who figures out how to make a big bang machine, to hitting the button. And it happens every 14 billion years. And that is the birth and the death of the universe: infinity." —Joe Rogan, comedian and host of *The Joe Rogan Experience*

"We don't know if there [already] was an anti-big bang, like the 'big squish' or the 'big squeeze.'"

—DR. NEIL DEGRASSE TYSON, ASTRO-SQUISH-ICIST

Let There Be Light

Will the Universe End in Fire or Ice?

"Some say the world will end in fire, / Some say in ice." Poet Robert Frost was talking metaphorically about love and hate—but we can certainly get astrophysical and apply his words to the end of the universe, too. "Take the temperature of the universe—they call it the cosmic microwave background," explains astro-lights-out-ist Dr. Neil deGrasse Tyson. "You can actually stick a thermometer in it and get a reading. It's about three degrees above absolute zero, and we're about 14 billion years old. When we're 28 billion years old, we'll be about one and a half degrees, and it'll scale right on down until we asymptotically approach zero.

"When you move metals you can end up generating currents, and if you have a current, you also have a magnetic field."

—DR. NEIL DEGRASSE TYSON, ASTRO-MAGNET-ICIST

"The temperature of the universe will continue to get cooler and cooler and cooler. Stars will eventually run out of their fuel. They will die. Then all matter will be left in the remnants, the dead cold remnants of stars. Once these energy sources run out, the stars will turn off one by one, the galaxies will shut off, and the universe will turn dark for the remainder of eternity."

So that's it, lights out? At least we're billions of years away from utter darkness. ∎

TOUR GUIDE

What's First: the Expansion of the Sun or the Loss of Our Magnetic Field?

Earth's magnetic field strength could drop drastically if internal motions, called convection currents, stopped in Earth's iron-nickel core. Current research suggests they might—a giant meteor impact could do it—but they would eventually resume.

Meanwhile, we know that the Sun will transform into a red giant and engulf our planet in about five billion years. So which comes first? "We'll probably lose our dynamo before the death of the Earth," says Neil. "'Dynamo' is lingo for the movement of molten iron."

Answer, an educated guess: The magnetic field goes first.

THINK ON THIS ▶ Can We Stop the Sun From Burning Out?

Nope. The Sun produces more energy in a thousandth of a second than the total of all the energy ever generated by humans in the history of civilization. Astrophysically speaking, it's possible to keep nuclear fusion in a star going by adding mass to it—perhaps when another star falls into it and merges with it. When we humans can make that happen, we'll be so advanced we won't need the Sun to survive anyway.

Can We Avoid the Fate of the Dinosaurs?

Dr. Hubbard knows what he's talking about. He's run NASA's Mars program, founded the NASA Astrobiology Institute, and is the architect of the Sentinel mission to detect killer asteroids. He's not insulting the dinosaurs, mind you: They dominated Earth for more than 150 million years. (We humans haven't even been here 5 million.) Then, 65 million years ago, a meteor 10 miles wide crashed into Earth's surface. It was the evolutionary blow that sealed their fate.

Dinosaurs hadn't evolved far enough to build tools and machinery. But humans have, and so today we can at least find killer meteors before they hit us. If we find one, what can we do?

"The reason that dinosaurs went extinct is because they didn't have a space program."
—DR. G. SCOTT HUBBARD, ASTRONAUTICIST AND AERONAUTICIST

Astrophysicist Dr. Amy Mainzer says we could smack it. "There are options to deflect it simply by whacking into it. Just hit it. The Deep Impact mission actually did that with a comet in 2005—just ran into the comet. And that actually can push something, if you have enough time."

Or we could paint it. You read that right. Neil shows us his artistic side and tells us how: "Because asteroids are dark, if you paint it white, it will reflect sunlight, and the sunlight bouncing off of that side will serve as a kind of mild propulsion to push it out of harm's way. But the problem is, it's not a big push; it's a gentle push." ■

Asteroids rain down on the last days of the dinosaurs.

Cosmic Queries: Mayan Apocalypse and Other Disasters

On the Scale of the Universe, Does Size Really Matter?

"In the year 1900 if you asked people what they were most worried about, it was overpopulation, and lack of food, and all this. They were not worried about asteroids. In a century, what will people be listing as the biggest risk to their lives? We have no idea. That's why it's good to learn what's out there."

—DR. NEIL DEGRASSE TYSON

When it comes to objects hitting Earth, you better believe size matters. A lot of stuff is continuously striking our planet, and we barely notice any of it, thanks to their small size. "Earth plows through several hundred tons of meteors a day. Some of that stuff falls in the day-

"Apophis is my favorite asteroid out there. It's the asteroid that's the size of the Rose Bowl that's headed our way. And it's not going to hit us on April 13, 2029 (which, by the way, is a Friday, in case you were wondering)."

—DR. NEIL DEGRASSE TYSON

time—you don't see it because it was bright," Neil explains. "And then, at night, are you looking up? Was it cloudy? Was it overcast? So you don't catch all of it when it falls. Most of it burns up, as what we call a meteor streak. But some of it is big enough to land—and then the meteor becomes a meteorite."

So, yes, in this case size really, really matters. If a single object the size of an apartment building were to hit us, it could flatten a whole city. But if that object were a mile across, all of human civilization would be at risk. ∎

THINK ON THIS ▶ Can We Just Send Bruce Willis?

In the movie *Armageddon* oil-workers-turned-astronauts zoom out to direct an asteroid away from Earth with a huge nuclear blast. Is there any reality to that?

DR. AMY MAINZER: If you have enough time, you might be able to do something that's just a simple kinetic impactor.

CHUCK NICE: Just like your health, early detection is the key.

"Halley's comet, in 1910, was the first discovery of cyanogen in the tail of a comet. At the time, we knew that Earth would be passing through the tail of the comet and people said, 'We're all gonna die.'"

—DR. NEIL DEGRASSE TYSON

StarTalk Live! From SF Sketchfest 2015

What Exploded Over Chelyabinsk, Russia?

In February 2013 a comet or asteroid estimated to measure about 50 feet across came in over Earth's north polar region. Fortunately, it exploded at high altitude; had it reached the ground before detonating, every tree and building in an area the size of Chicago would have been flattened. Sentinel mission architect Dr. G. Scott Hubbard explains how it could have been worse: "It damaged about a thousand homes. It blew up at 60,000 feet. It was like an airburst coming down. About 1,100 people were sent to the hospital, but nobody died . . . If you're talking about the size of an asteroid that's in the miles, then you're talking extinction event, as in what happened to the dinosaurs . . . Then, if you get down to the 100-meter-ish—300-feet-or-so—range, you're talking about cataclysms that are city killers. And then, down to the 30- to 50-meter range, you have tsunamis . . . There are maybe a million of these near-Earth objects out there." ∎

A fragment punched through the ice of Lake Chebarkul

What Happens When a Nuclear Bomb Detonates?

At least one doomsday scenario is real: If a significant fraction of the 20,000 or so nuclear bombs in the world's arsenals were detonated, the energy released would approach that of a one-mile-wide asteroid striking Earth. Until the 1990s, when the Cold War ended, people lay awake at night worrying about such a scenario. We may not worry as much today, but the possibility is still out there, and real.

> "You get the air, the expanding air, which then blows everything into the winds. All we are is dust in the wind."
>
> —DR. NEIL DEGRASSE TYSON

Nuclear warheads harness the power of fission, as large atoms like uranium and plutonium are broken apart, or fusion, as small atoms like hydrogen are combined. These processes convert matter into energy, which comes out as light in many wavelengths—from gamma rays to radio waves—and kinetic energy, in the form of shock waves and winds. Fission and fusion also produce vast amounts of radioactive by-products that can survive for years, even centuries, after a nuclear warhead has been detonated. These by-products can create long-lasting health hazards. ■

An atomic explosion and shock wave

> "You get the light, and it is hugely intense, and it basically vaporizes you. It will melt you. It'll vaporize you. It'll burn you. Then, the shock wave, which is moving at the speed of sound, breaks everything apart."
>
> —DR. NEIL DEGRASSE TYSON

LAUGH OUT LOUD ▶ **With Neil and Chuck Nice, Comedian**

NEIL: I'm more afraid of zombies than dinosaurs. The big dinosaurs with the big teeth—they did not survive the asteroid.

CHUCK: So they are living underground. They're just oil now.

NEIL: Soon to exact their revenge on our ecosystem by increasing our carbon footprint. The revenge of the dinosaurs. Actually, our oil is mostly vegetation.

The Future of Humanity With Elon Musk

Should We Be Afraid of Artificial Intelligence?

Opinions differ widely about the benefits or dangers of artificially created sentience, from Wall-E to Skynet. Here are two divergent views, from Elon Musk, founder of Tesla Motors and SpaceX, and Bill Nye the AI Guy:

"I'm quite worried about artificial superintelligence these days," says Musk. "I think it's . . . maybe something more dangerous than nuclear weapons . . . What does it try to optimize? And we need to be really careful with saying, 'Oh, how about human happiness?' Because it may conclude that all unhappy humans should be terminated, or that we should all just be captured and [have] dopamine and serotonin directly injected into our brains . . . We should exercise caution."

Nye is more optimistic: "I'm all for the singularity when computers are as smart as people. But computers, and the quantum-computing thing, run on electricity . . . The robots show up in western China and there's no place to plug in . . . They're not very productive, let alone take-over-the-world-ive. There's a lot of other things to worry about."

Artificially intelligent beings will probably wind up exactly like natural intelligent beings. Which ones will be more dangerous? ■

TOUR GUIDE

Is IBM's Watson Intelligent?

Watson was programmed to be a contestant on the quiz show *Jeopardy!*—and soundly defeated its human competition. "I would call [Watson] intelligent, but I wouldn't call it 'near-human,'" says Stephen Gorevan, chairman of Honeybee Robotics. "There are degrees of accumulated knowledge and organized ability, and I think that when you get . . . into the realm of art, you'd see the difference."

Who defines intelligence, anyway? Neil and Stephen are both knowledgeable and intelligent, after all—and Watson could beat either of them at *Jeopardy!* Maybe Watson wouldn't call them "near-Watson."

Designing AI requires that we define human intelligence and emotion.

"If there was a very deep digital superintelligence that was created that could go into rapid recursive self-improvement in a nonlogarithmic way, then . . . it could reprogram itself to be smarter and iterate very quickly, and do that 24 hours a day on millions of computers. Well . . . that's all she wrote . . . We would be like a pet Labrador—if we're lucky."

—ELON MUSK, FOUNDER OF TESLA MOTORS AND SPACEX

Cosmic Queries: Human Impact on Earth With Dr. FunkySpoon

Who's More Dangerous: Humans or Cyanobacteria?

Cyanobacteria—also known as blue-green algae—were the first Earth organisms to use photosynthesis, which combines carbon dioxide and water to make sugar compounds for food. "The cyanobacteria, 2.2 billion years ago, evolved photosynthesis, and they thought, 'Here's a great energy source, sunlight; this is wonderful,'" explains astrobiologist Dr. David Grinspoon. "And they started polluting the air with oxygen, which caused a catastrophe and wiped out most of the species that were alive then."

"So interestingly, we're not the first species to come along and, in the quest for an energy source, screw up the planet."

—DR. DAVID GRINSPOON, ASTROBIOLOGIST

Thanks to the presence of cyanobacteria, oxygen gas—the waste product of photosynthesis—eventually supplanted carbon dioxide in Earth's atmosphere. It took a long time for primitive life on Earth to evolve to use that oxygen, which is still deadly to a lot of modern-day microbes. But it also took the cyanobacteria a very long time to poison the air.

Compare the effects of cyanobacteria with the climate change we are seeing today—rapid changes to the Earth's atmosphere from greenhouse gases. For better or worse, we humans seem to be doing in just a few centuries what it took cyanobacteria eons to achieve. ∎

Microorganisms find strength in numbers.

DID YOU KNOW

There are an estimated 2,000 to 8,000 species of cyanobacteria, and they can survive in almost every kind of ecosystem on Earth.

LAUGH OUT LOUD ▶ **With Michael Che, Comedian**

"I heard that the cows fart and it's ruining everything, but it seems like a hilarious way to go . . . What if it was written in . . . like an ancient book, that someday the cows will all fart at the same time, and that will be the end of mankind? . . . But it's stranger than fiction."

Cosmic Queries: Mayan Apocalypse and Other Disasters

Could a Doomsday Solar Flare Wipe Out Life on Earth?

Solar flares regularly burst forth from the sun and spray radiation and solar particles into the solar system. A solar flare is unlikely to cause our extinction, but it might be strong enough to create a serious electrical headache. "If this flare is particularly potent, charged particles can reach lower in our atmosphere and affect our communications satellites," says Neil. "Our satellites run on electrical currents. If you have charged particles swarming electrical devices, you can create short circuits. A really big flare could sort of leave us . . . communications-blind. And the electrical grid could also be affected, because if a flare gets really low it can reach the Earth and short-circuit it. We've got top people working on it, but we are susceptible." ∎

"If the global climate changes in ways that our systems cannot respond to, it would mean widespread death, given how we now are dependent on the equilibrium and stability of the system that we created for ourselves."

—DR. NEIL DEGRASSE TYSON

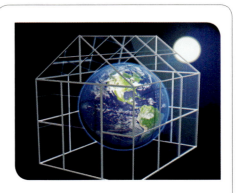

TOUR GUIDE

Could We Suffer a Runaway Greenhouse Effect, Like Venus?

Research shows Venus once had oceans, but they all boiled away—a result of a runaway greenhouse effect that ratcheted up its surface temperature to 900°F. "If we did the worst-case scenario—burned all of our coal, all of the tar sands, fracked the Earth to pieces—could we push Earth to a Venus-like state?" astrobiologist Dr. David Grinspoon asks. "We actually don't know; there are different opinions about this. In some sense, it's academic, because even if we didn't push Earth literally into runaway greenhouse, where the oceans went away . . . we would push Earth easily beyond the point where we could live."

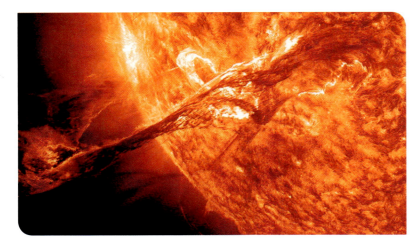

Solar flares can extend thousands of miles above the Sun's surface.

"I want to stand up for the human race . . . We're having this conversation; we're becoming aware of our role on the Earth. So I actually think that makes us different . . . We're slime, but we're smart."

—DR. DAVID GRINSPOON, ASTROBIOLOGIST

BEING HUMAN

We build. We create. We've walked on the moon. We've constructed communities and vast cities and great monuments. We laugh, sing, play, eat, drink, and love. Being human is awesome! We also demolish. We destroy. We've wiped out ecosystems. We've waged wars, leveled cities, and committed terrible violence. We cry, scream, fight, puke, choke, and hate. Being human is scary! How did humanity get this way? How will we change in the future? Those questions have answers only if a human being can know the meaning of being human.

"It's why we all exist. It's why all living things exist. It is an astonishing fact that on this planet the laws of physics got channeled through this remarkable process, Darwin's evolution by natural selection, to produce objects like us."

—DR. RICHARD DAWKINS, EVOLUTIONARY BIOLOGIST

CHAPTER ONE

If We Evolved From Monkeys, Why Are Monkeys Still Here?

Few theories in science are as controversial—and as misunderstood—as evolution by natural selection. The process doesn't make us better; it makes us different. And it does so reluctantly, in fits and starts, and agonizingly slowly. It can work really well, and keep a living organism around for a really long time, and then—boom! Its inability to keep that same organism alive is painfully displayed as yet another species bites the dust.

  Yes, humans evolved from monkeys. But whether or not you agree, evolutionarily speaking we're not better than monkeys; we're different. There are still monkeys because they evolved to live in trees (where they sure are better off than we would be). We would all do well to remember this fundamental tenet of evolution. It would help us understand who we are, what we are, and where we're headed—whether it's into space or in front of the TV.

Humans and gorillas split from a common ancestor millions of years ago.

StarTalk Live!: Evolution With Richard Dawkins (Part 1)

What Is Darwin's Theory of Natural Selection?

Evolutionary biologist Richard Dawkins uses computers to explain reproduction and evolution: "There is highly accurate, high-fidelity information, really like a computer language, which is copied from generation to generation, and within each generation programs the development of the body in which it sits. Therefore, the fate of the program is bound up in the fate of the body in which it sits."

In every living organism on Earth, just such a computer program runs, coded with the creature's DNA. If the program is successful, the organism can reproduce, and the program survives in the offspring. If the original program isn't perfectly copied, the new organisms are different. If that change helps those new organisms reproduce, then those changed programs survive in the next generation. The process repeats—until, thousands of generations later, the result is the amazing variations of life found on Earth today. ∎

One of Darwin's finches, unique to the Galápagos Islands

BACK TO BASICS

David Attenborough Explains the Mystery of Life

"You can chronicle the history of life in a surprisingly detailed way in quite a short period. You know that life starts in the deep sea, and it leads to different types of invertebrates and shell, and crustaceans and shrimps and so on. Then there are fish with backbones, and fish with backbones emerge onto land and become amphibians with wet skins, and amphibians with wet skins get dry skins and become reptiles. And some of the reptiles turn their scaly skins into feathers and become birds, and the others turn them into hairs and become mammals . . . That's what the history is, and you can put as much or as little detail on that skeleton as you like."—Sir David Attenborough, naturalist

"We've seen it, in a human lifetime: a new species of insect come into being. You make predictions with theories, and here it happens."

—BILL NYE THE EVOLUTION GUY, ON THE LONDON UNDERGROUND MOSQUITO

Cosmic Queries: Planet Earth

Has Gravity Affected Evolution on Earth?

Earth's gravity played a crucial role in setting the stage for life as we know it. It allowed our planet to hold on to an atmosphere that produces enough pressure to keep water liquid at its surface. Its downward acceleration caused many important substances on Earth—especially rock, carbon, and water—to cycle up and down near the surface, mixing the raw ingredients crucial for the dawn of life. The Moon's gravity mattered, too; many biologists hypothesize that tides created ecosystems that continually changed from wet to dry, fostering the evolution of land animals as they transitioned from living in water.

Isaac Newton contemplates gravity.

That said, many lifeforms aren't affected much by the force of gravity in their daily existence. For example, bacteria moving around in a drop of pond water are more tightly bound by the water's surface tension; their masses are small, and they are neutrally buoyant. Most life, though, does feel gravity's influence profoundly—and generally speaking, the more massive we are, the more we're affected. ∎

CONVERSATION

Why Don't We Have Wings?

NEIL
Having a feature doesn't always mean you'll be better at reproducing.

RICHARD DAWKINS
If a man of our size had bird-sized wings, he couldn't get off the ground.

EUGENE MIRMAN
Why couldn't you just have enormous wings that were fairy-sized in proportion?

BILL NYE
We are constrained by the same laws of physics.

"Why don't people have wings? They wouldn't be better off with wings. Wings can get in the way."
—RICHARD DAWKINS, EVOLUTIONARY BIOLOGIST

THINK ON THIS ▶ **Is Evolution a Miracle of Creation?**

"I'm a firm believer in evolution . . . I don't understand people who can't believe in evolution . . . I have a hard time understanding why people cannot accept the fact that God can work through evolution, that it's just as much a miracle of creation if it takes 10 million years or 15 million years as if it took seven days." —Rev. James Martin, Society of Jesus

Evolution on Earth

Looking at the complexity of life on Earth today, it's tempting to believe that something or someone had to plan it all out. But there was no need; just put together enough DNA, environmental conditions, and time, and a set of environmental stressors—and it all happens naturally.

Eyes

◄ The evolutionary path from clusters of photosensitive cells to organs that produce detailed images has happened independently in the animal kingdom dozens of times. Octopus eyes, for example, don't have the "blind spot" human eyes have, because their optic nerve fibers are located behind the retina.

Echolocation

◄ Dolphins are swimming mammals, and bats are flying ones—yet each developed the ability to hunt prey by emitting high-pitched sounds and listening to their echoes. Research suggests that they each evolved that ability independently but based on the same genetic mutations.

Wolf to Pomeranian

◄ Thousands of generations of evolution and breeding have transformed wolves from a frightening and fierce predator into the full spectrum of domesticated dogs—including cute and cuddly little toy varieties.

Stingers

▲ Female insects often deposit their eggs using a penetrating appendage in their abdomens. Worker bees, which are all female, have a similar organ—but it deposits venom as a weapon in defense of its hive's egg layer, the queen.

Giraffes

◀ Unlike apatosaurs or barosaurs, which evolved more and longer vertebrae, giraffes have the same number of neck bones as humans. Variations in those neck bones led to both short-necked and long-necked giraffes over millions of years.

Primates

◀ The first protoprimates appeared on Earth just as the dinosaurs went extinct. About 15 million years ago, the great apes began their evolutionary fragmentation, splitting into the four types that survive today: orangutans, gorillas, chimps, and us.

Homo sapiens versus Homo neanderthalis

▲ For years, we thought Neanderthals were our ancient ancestors (most of us have some Neanderthal DNA)—until scientific testing showed they most likely represent a different branch of our hominid family.

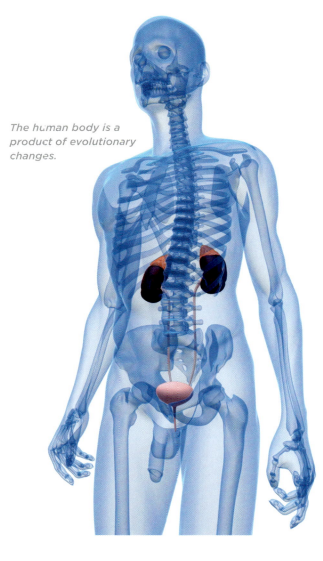

The human body is a product of evolutionary changes.

DID YOU KNOW

The human genome contains about 20,000 genes. We have very little idea about what most of them do.

What's Intelligent About This Design?

Some people refuse to accept that evolution by natural selection occurs—and some of those people support the hypothesis called intelligent design. The complexity and beauty of life is so amazing, they assert, that some agent or agents with sentience and awareness must have intentionally created everything. Evolution is impossible, because nothing as complex as human life could possibly have happened by accident.

Oh, really? Are our bodies well designed? Well, actually, some parts of us seem a little stupid, according to Neil: "We have a human body where, between our legs, we have a sewage system in the middle of an entertainment complex. No engineer would have designed that, I'm certain."

"Plus the fuel tank—the nozzle for the fuel tank—is right next to the air filter," Bill Nye the Science Guy agrees.

That's right, says Neil. "So some percentage of us choke to death. We're alive in spite of our body's design." ■

> "An utterly natural process gave rise to this magnificent complexity, this beauty, and this illusion of design. It works by very, very slow change from generation to generation."
>
> —DR. RICHARD DAWKINS, EVOLUTIONARY BIOLOGIST

LAUGH OUT LOUD ▸ **With Bill Nye and Comedian Jim Gaffigan**

JIM: If there's genes, why do I have so many recessive genes? Balding. Eyesight. Pale, sexy skin?

BILL: One of the remarkable insights in evolution . . . is you only have to be good enough. I'm not just talking about you. I'm talking about all of us.

Planet of the Apes

How Similar Are Humans to Apes?

Humans, orangutans, gorillas, and chimpanzees all evolved from a common genetic ancestor 15 million years ago. Paleoanthropologist Ian Tattersal explains how our behavior and thinking compares to these apes: "Apes can recognize themselves in mirrors . . . They have a sense of fairness . . . They definitely have attachment. They definitely have feelings of dislike. They have positive and negative feelings."

Take a look at these three chimp behaviors; do they remind you of anyone you know? ■

TOUR GUIDE

How Does Andy Serkis Play Chimps?

Andy Serkis is one of the world's leading motion-capture actors—meaning that sensors translate his movements into computer-generated graphics. Appearing in the recent Planet of the Apes movies, he devolved into one of our evolutionary ancestors thanks to technology that he says has evolved a lot: "In *King Kong* days, I had 132 tiny spherical three-dimensional markers stuck all over my face, including my eyelids . . . Facial performance capture is totally evolving, as is the use of head-mounted cameras for real high-definition video reference, with markers all over your face—that's where it's at [now]. But I think there will be a version without that, where it's purely optical and we will have done away with all that in the near future."

A chimp looking relaxed, with mouth hanging slightly open, indicates general satisfaction and comfort with its surroundings.

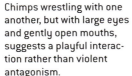

Chimps wrestling with one another, but with large eyes and gently open mouths, suggests a playful interaction rather than violent antagonism.

When a chimp bares its teeth broadly and narrows its eyes into small beads, watch out—it is most definitely not happy to see you.

"I was thinking, I wanted to create creatures just human enough, so that when they did things like humans did, it was very funny. And so that's why I created monkeys . . . That's the basic principle behind the entire line of primates. Monkey funny. Chimpanzee funny. Gorilla funny."

—@THETWEETOFGOD

LAUGH OUT LOUD ▶ **With Kristen Schaal**

"Bonobos are incredible . . . They'll hump you to say hello. They're amazing . . . We went into the jungle and we lived with the bonobos so we'd learn their language from them. [Sex is] their vehicle to communicate."

Pacifist bonobos bond through grooming.

Can't We All Just Get Along?

Neanderthals may have battled with our Homo sapiens *ancestors. If so, they lost.*

War ranks as perhaps the worst group activity humans have invented. But did we invent war? "It is certainly documented that sometimes bands of males organize themselves, go into the territories of other males, with apparently the ultimate aim of taking over those territories," explains paleoanthropologist Dr. Ian Tattersall. "They draw blood. They murder each other . . . They're not good at throwing things, but they can whack each other with sticks."

We can go even further down the evolutionary chain if we want. Ant colonies wage massive wars; some species even enslave the workers they conquer, forcing them to serve their victors. Sometimes those slave ants revolt, killing the pupae that they were forced to care for.

On the bright side, bonobos get along remarkably well, famously engaging in group eating, sleeping, and frequent sexual activity. And they belong to the chimpanzee genus, which means they're genetically closer to us humans than any other animals. ∎

Cosmic Queries: Primate Evolution

Are We More Like Bananas or Chimps?

More than half of your DNA is identical to the DNA in a banana. Humans and mushrooms are genetically more alike with one another than either is genetically akin to any green plants. In fact, we share pretty big proportions of genetic material with many different organisms. ■

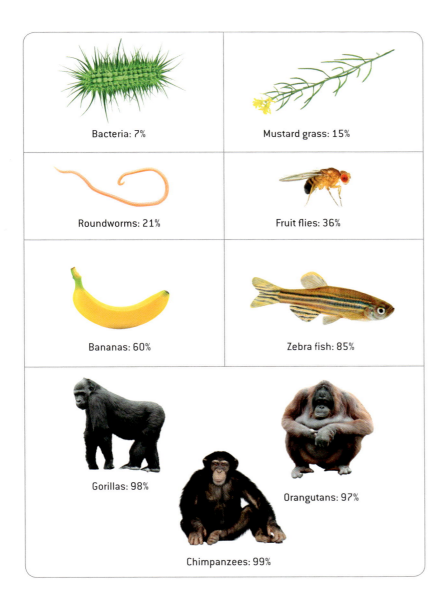

Bacteria: 7%

Mustard grass: 15%

Roundworms: 21%

Fruit flies: 36%

Bananas: 60%

Zebra fish: 85%

Gorillas: 98%

Orangutans: 97%

Chimpanzees: 99%

BIOGRAPHY
👓

Why Was Charlie D. So Awesome?

Born to a wealthy and well-educated English family, Charles Darwin (1809–1882) dutifully followed his parents' wishes to study medicine and religion. Science was his calling, though, and after college, he took the opportunity to sail around the world for nearly five years as ship's naturalist aboard the H.M.S. *Beagle*. His travels, collections, and writing made him a respected naturalist by the time he came home to England. Darwin developed his theory of evolution for 18 years without publishing it, knowing the controversy it would create. Finally a younger colleague, Alfred Russel Wallace, announced he had come up with a similar theory. The two presented their work at the same time, and a year later Darwin published his seminal book, *On the Origin of Species by Means of Natural Selection*.

"Really, he just was jammin', looking at his collections and reaching amazing conclusions, world-changing conclusions. Gotta respect that."

—BILL NYE THE JAMMIN' GUY

Is Technology Helping or Hindering Human Evolution?

Evolution by natural selection as originally conceived by Charles Darwin was a fledgling theory. Even though it got the fundamentals right, Darwin couldn't answer every question he had—which is why scientists have been challenging, testing, and refining the theory of evolution ever since.

> Bill Nye recounts how, thanks to medical technology, he and his DNA are still able to contribute to human evolution: "A guy took out my appendix; I'm still going. I could have babies and things."

Now another wrinkle to the theory has come about. Thanks to advances in everything from medicine to agriculture to computers, individual humans who would not have survived even a few years ago can now lead full lives and pass their DNA along. By Darwin's definition, this "artificial" technology truly amounts to "natural" selection—the means by which a species can adapt itself to its environment, and now not just over hundreds of generations but in less than one.

Is our rapidly increasing ability to survive and reproduce stunting our biological evolution? We'll have to wait for many more generations of humanity to find out for sure. For now, though, humans are increasing rapidly in both population and life expectancy—which, biologically, implies evolutionary success. ■

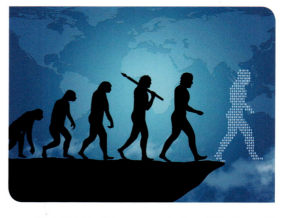

The stages of human evolution. What's next?

DR. IAN TATTERSALL:
There are lots of drugs out there that will damage your DNA.
There are no drugs available that will [increase intelligence].

EUGENE MIRMAN:
So it's easy to damage DNA but hard to become superhuman? Oh well.
Can't have everything.

THINK ON THIS ▶ Are We Evolving to Be Post-Human?

"Certainly we're going to be tinkering with our biology . . . When biology becomes an information technology, we master those information processes of biology and we can reprogram, basically, with the canvas of life itself. We can upgrade . . . We can turn ourselves into something that's post-human, [which is] much more interesting and maybe not subjugated to the same limitations that we currently have." —Jason Silva, futurist

Packing for Mars (Part 2)

Could a New Human Species Appear?

New species are continually being created. Just isolate a breeding population of an existing species in a new environment, then wait for natural selection to create genetic variations. "If you create a colony on Mars, and that colony does not crossbreed with anyone on Earth for a thousand generations, then there's the risk (or perhaps maybe that's what you want) of their becoming another kind of species, with different properties that enable their survival in that location," Neil explains. "That would be kind of interesting."

Interesting, but problematic—at least according to Chuck Nice: "It would be, especially when they return to Earth to conquer us all, because you know that's what's gonna happen."

"That would be bad," agrees Dr. Mike Massimino. "Don't give them any return directions. Just send them out there and tell them to send video." ■

TOUR GUIDE

Can We Breed Men to Lactate?

Comedian Maeve Higgins was once told a remarkable story by her mother in which a ship wrecked off the coast of Japan. Only men and a baby survived. On the life raft, one of the men was able to lactate to feed the baby.

Could this story be true, could a male really be able to produce milk? After all, men do have nipples.

"You could probably make males lactate by giving them hormone injections," says evolutionary biologist Dr. Richard Dawkins. "You use the hormone injection as a way of bringing to the surface genetic variation which was there all the time."

In the sci-fi drama The Expanse, *a human colony on Mars plots war against Earth.*

LAUGH OUT LOUD ▶ With Comedian Jim Gaffigan

Neil and several guests were discussing what could be achieved if humans were to be genetically modified. Everyone agreed that people might be made who could become brilliant musicians. "But for every musician you tried to make, you'd make a lot of baristas, right?" says Jim Gaffigan.

Rocket City Rednecks

Are We Getting Smaller and Stupider?

Over the course of the 20th century, average IQ has been going up massively," says Dr. Richard Dawkins, evolutionary biologist. "And I must say, I don't notice it, but the data seem to be there, and I wish it were true."

In the past century, a wide variety of measures have suggested that, on average, humans in the developed world have been getting both taller and more intelligent. Some hypotheses on the causes of these trends include better nutrition, less disease, more education, and a more stimulating life environment overall.

Will the current state of humanity continue the trend? We're squinting at electronic screens, hunched over at our desks, eating junk food, and cutting recess from our kids' school days. For better or worse, we'll find out only a generation from now. ∎

TOUR GUIDE

Do You Want Fries With Your Ph.D.?

The next time you're tempted to judge a person's intelligence by a stereotype, think about this fun fact from Dr. Travis Taylor, engineer and host of *Rocket City Rednecks*: "North Alabama, particularly around the Huntsville-Decatur area, the Rocket City area, has the highest average IQ of any place in the country. . . When I was working on my first Ph.D. in physics, there was a girl in the department . . . [who was] working on her Ph.D. in astrophysics, and she paid her bills by working at Hooters. The waiters and waitresses at the time—they were going through school. There was no telling what you would be talking about in a restaurant—how you're going to solve this problem or whatever—and somebody else in the room would have some input . . . even the Hooters waitress."

Is this the face of a generation with climbing IQs?

Cosmic Queries: GMOs With Bill Nye (Part 1)

Is Genetically Modifying Organisms the Same as Evolving Them?

Humans have been genetically modifying organisms for millennia. Every food item from the supermarket comes from a GMO—and the modification arose from various experiments, starting with seed selection in ancient times and continuing to a planting bed or a livestock bin or an indoor laboratory.

Thanks to human activity, the environment has been changing faster than at any time in recorded history. Changing plant life on the timescale of a few generations, rather than hundreds or thousands, may be part of how we adapt to those changes—including

All supermarket apples have been genetically modified.

how to feed, clothe, and shelter us into the future. "I met the guy who won the World Food Prize," Bill Nye the Sci-Food Guy says. "It's like the Nobel Prize for agriculture . . . And he believes that we can raise more food than ever on less land . . . He and his colleagues believe they can raise food for nine billion people on 2 percent less land. That's a noble goal." ∎

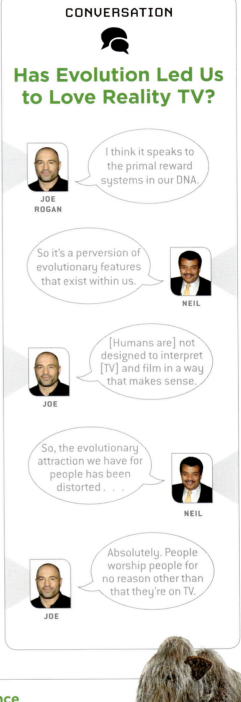

CONVERSATION

Has Evolution Led Us to Love Reality TV?

JOE ROGAN: I think it speaks to the primal reward systems in our DNA.

NEIL: So it's a perversion of evolutionary features that exist within us.

JOE: [Humans are] not designed to interpret [TV] and film in a way that makes sense.

NEIL: So, the evolutionary attraction we have for people has been distorted . . .

JOE: Absolutely. People worship people for no reason other than that they're on TV.

LAUGH OUT LOUD ▶ With Sally Le Page, GE Creator in Residence

When asked what extinct species they'd bring back, Bill Nye chose a woolly mammoth or flying raptor while Neil went for a pet dinosaur—"a lap *T. rex*," to be exact. "So, in our scientific Jurassic World, we would have both demonic flying dinosaurs and feathery, cuddly pet dinosaurs," Sally Le Page clarifies. "Really soft, fluffy dinosaurs."

"I think there should be more sex in labs —and I'm sure it's happening, like, you know, late nights, anyways."
—KRISTEN SCHAAL, CO-AUTHOR OF *THE SEXY BOOK OF SEXY SEX*

CHAPTER TWO

Can Science Help Me Find True Love?

On bulletin boards everywhere, flyers are posted with the big letters *S-E-X* across the page. In much smaller type below, something like this is written: "Now that we have your attention, please attend next week's meeting of the crochet club at the community center." It works every time. Why?

Without reproduction, we humans wouldn't be around. Somewhere along the way, though, the biology of sex got infused with much more. The social and psychological fabric of our entire species seems tied to the success—or failure—of achieving the related goals of love and sex, whether or not reproduction actually happens.

Romance and reproduction carry so many strong social conventions, instructions, and prohibitions that talking openly is usually taboo in polite company. But not here! Respectfully but fearlessly, scientists and nonscientists alike study and consider what makes us want to love, what happens to our brains and bodies when we love, and how love and sex show themselves.

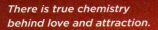

*There is true chemistry
behind love and attraction.*

The Evolution of Love and Sex, With Dan Savage

Is Love at First Sight for Real?

Physical attraction between two people can happen almost instantaneously; psychological or emotional attachment takes longer. If love is a combination of all those kinds of attraction, as current research suggests, the "first sight" phenomenon is probably a flash of infatuation or lust that blossoms into love as that first instinct is explored further.

Biologically, it is completely possible for two people to connect and stay permanently connected emotionally after their first meeting. What makes that attraction happen, though, is still mysterious. All kinds of factors—including height, weight, facial features, arm length, and even body odor—have been tested, with mixed results. "It's very easy to explain love at first sight," says biological anthropologist Dr. Helen Fisher. "The brain circuitry for romantic love is like the fear system. You can be scared instantly, and you can fall in love instantly." ∎

The cause of a "spark" still eludes scientists.

CONVERSATION

What Gets Us Excited?

JOSH GROBAN
I mean, [Neil starts] talking about multiverses, and the pants just drop.

BILL NYE
Without sex, your genes wouldn't live long at all.

DR. RUTH WESTHEIMER
Anything consenting adults do, on the kitchen floor, living room couch, bedroom, is perfectly all right.

DR. HELEN FISHER:
I put people in brain scanners and study the brain circuitry of romantic love.

CHUCK NICE:
Talk about your foreplay.

THINK ON THIS ▶ Is Love a Drug?

Love causes the production of all kinds of hormones and chemicals—like dopamine, serotonin, oxytocin, and vasopressin—that cause your brain and body to do all kinds of things. Drugs do that, too. What's more powerful? "Sex is a drug, definitely—a huge drug. But an even bigger drug is romantic love," explains Dr. Helen Fisher, a biological anthropologist. "You ask somebody to casually go to bed with you and they say, 'No, thank you,' you don't kill yourself."

The Evolution of Love and Sex, with Dan Savage

Are Humans Monogamous—or "Monogamish"?

Humans have been around a long time, and acceptable social behavior has continually changed. Current mainstream American society tends to praise exclusive one-on-one sexual relationships. But does this come from nature? The scientific research on the subject is inconclusive. Let advice columnist Dan Savage explain: "What we know about primates and mammals [is that] we are not a naturally monogamous species. We are a pair-bonding species. But there's social monogamy, which is the pair bond, and there is sexual monogamy, which is never touching anybody ever again with your genitals . . . No primates with testicles our size are monogamous sexually . . . It is a difficult struggle for us."

But there may be a fix. And Dan Savage has a simple solution: Change the nature of relationships to better reflect the desires of humans who want to be sort of monogamous, to have what he calls "monogamish relationships."

Is this just a cute name for a phenomenon that goes on already throughout human society? Maybe. Subcultures certainly already exist today that allow their members to behave "monogamishly" with little social consequence. Sometimes, though, it can have unexpected biological results. In Swaziland, for example, where husbands routinely have numerous sexual partners while married, the HIV/AIDS infection rate is extremely, and depressingly, high. ∎

Marrying for love is a relatively new concept in human history.

"There are no high-maintenance items in my house of any kind: pets, plants, or husbands."

—DR. CAROLYN PORCO, PLANETARY SCIENTIST, TO A REPORTER ASKING HER WHY SHE WAS STILL SINGLE

The Science of Sex [Part 2]

How Do Scientists Study Arousal in the Lab?

Arousal—and, frankly, just about every other aspect of human sexual activity—can be tough to study under controlled laboratory conditions because of the social taboos surrounding them. Even if human test subjects were willing to be observed, poked, and prodded for such things, how can we know if the subjects' responses are natural? Maybe it's a matter of finding the right environment and volunteers—perhaps on the set of a pornographic movie.

Not so fast. "Actually, that's not a good idea," says Mary Roach, author of *Bonk*. "Masters and Johnson started out using sex professionals, but they found out they were too good at faking it . . . [The researchers] wanted someone more representative of the average person."

Nevertheless, scientists soldier on. A lot of it is about measuring physical responses—blood flow, pupil dilation, breathing, heart rates, and so on—to stimuli. Things are generally desensationalized, though; for example, pornographic material in sex labs is called "visual erotic stimulation." ■

"When a man ejaculates, 120 million sperm travel through the vas deferens at an average of 28 miles per hour . . . But once they're transported out of the penile shaft, they slow down to 0.0011 miles per hour."

—DR. NEIL DEGRASSE TYSON, ASTRO-SPERM-ICIST

Labs are not designed with mood lighting.

"[There's] this four-dimensional ultrasound moving-picture image . . . You can see a puckering lip or an erecting penis . . . It felt less like sex than some awkward thing you have done at the hospital . . . I took one for the reader."

—MARY ROACH, AUTHOR OF *BONK*, ON THE STRANGEST PLACE SHE'S EVER HAD SEX

THINK ON THIS ▶ Do You Have a Cold, or . . .

Are you just happy to see me? Erectile tissue, medically speaking, has a lot of vascular spaces that can become engorged with blood or other bodily fluid. The genital area isn't the body's only erectile tissue, though. Mary Roach, author of *Bonk*, explains the other: "The only other erectile tissue in the human body is in the nose. When you have a cold, you basically have a nose-boner . . . The nipples, that's a different erection system: That's muscles squeezing."

What Makes Music So Seductive?

Brain scientists know that hearing depends on an extremely complex system, from tiny vibrations in the air moving toward your eardrums to a vast sensory and analytical network in your brain. Current research shows that sound affects every part of your brain, from its primitive core to its highest executive functions. Music in particular seems to trigger a host of brain responses involving recognition (whether you know a tune or it's similar to one you know) and pleasure (the release of dopamine and serotonin).

Here's what musician Moby had to say: "Music is so ubiquitous, it's such a normal part of our lives—but it can do so much. They play it at funerals. They play it at weddings. People play music to have sex. They play music to cry. People play music when you're trying to get armies to march in to war. And what's amazing about music to me, it doesn't exist. All it is, is air moving a little bit differently. But somehow, moving air a little bit differently can make someone weep, can make someone jump up and down, can make someone move across the country and cut their hair . . . I don't want to figure it out. I just love that it has this power." ■

"There's a wide-eyed wonderment that I have to where that [music] comes from, and what causes it."

—JOSH GROBAN, MUSICIAN

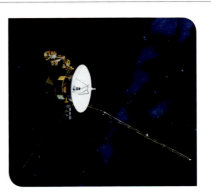

TOUR GUIDE

What Does Love Sound Like?

Crooners of romantic ballads know how to turn a melody with lyrics into human desire and passion. Usually, according to research, it's not about the precision with which the song is delivered; rather, it's the combination of familiar melodies with variations in tempo (musicians call it "rubato"), rhythm, and vocal tone that drives us wild.

How are those kinds of subtleties captured and recorded by the human brain? Could such recordings be recognized by others—aliens, even? Neil asked Ann Druyan, wife of famed astrophysicist Carl Sagan and producer of the *Cosmos* television series, what her favorite sound was on the "Golden Record" now traveling beyond the solar system on the Voyager spacecraft.

"Well, I have to say: . . . My brain waves, two days after Carl and I fell madly in love with each other. [They are] part of an hour-long meditation in which all of my neurological electrical impulses have been recorded, and that data translated into sound on the record, so that extraterrestrials of the distant future might be able to understand something of what it was like to be alive."

The Science of Sex [Part 1]

The Science of Sex [Part 1]

How Has Sex Changed Over the Years?

Socially speaking, sexual behaviors in much of the world have changed drastically in the past decades. The largest effects, according to researchers, have come from the availability of birth control and from increased amounts of leisure time. "First of all, it has changed because of work situations," explains sex therapist Dr. Ruth Westheimer. "You don't work anymore 12 hours like some people used to. You work eight hours . . . More time. The other thing is: No question that the pill has drastically changed the whole mechanism, the whole attitude, everything, in terms of women having control, and in terms of not having to worry about an unintended pregnancy . . . Some maybe not-so-good side effects of sex during lunch, not with your partner, or something like that, is also part of it."

Whether or not birth control has led to more sexual activity, one fascinating consequence of it appears to be a lower rate of pregnancy termination. In Uganda, for example, where abortion is illegal and sex education focuses on abstinence, the abortion rate is more than twice the rate in the United States and four times that of western Europe, where both abortion and contraception are legal and widely available. ■

TOUR GUIDE

Does Viagra Work for Women?

"There's a disconnect with women between the vagina—the body—and the brain. Viagra actually does work on women—it does increase vaginal blood flow—but the women don't pick up on it. They don't feel aroused. It does have an effect, but it doesn't do what it's supposed to do, which is make them feel like they want to have sex. It doesn't affect the libido. That is an interesting disconnect that isn't there with men so much."—Mary Roach, author of *Bonk*

"Women have nocturnal erections in the same sort of cycle as men. Little tiny clitoral erections. [To learn] that, somebody wired up a fairly large clitoris with a strain gauge, figured it out."

—MARY ROACH, AUTHOR OF *BONK*

THINK ON THIS ▶ Are Women Aroused by Porn?

"They found, surprisingly, that men were aroused by porn that fit their sexual orientation, but the women responded to anything, the entire spectrum. People think of men as being visually stimulated and women just could care less. If you asked the women, they might say, 'That didn't do anything for me.' But their vaginas were saying otherwise." —Mary Roach, author of *Bonk*

What's the Secret to a Long-Lasting Marriage?

Scientists who conducted a multiyear study on a large set of married couples showed that 93 percent of the couples who divorced engaged in four kinds of destructive behavior when they disagreed with one another: contempt, criticism, defensiveness, and stonewalling.

Experts generally agree, that these kinds of behaviors can lead to the end of a relationship. But they also offer advice for what behaviors do work. "I think being good to each other, taking care of each other, and not taking each other for granted. And to try to keep things in perspective," says advice columnist Dan Savage. "You have to identify those things that, no matter how much you bitch and complain about them, will never change . . . and not . . . complain about it and guilt-trip them about it all the time—put up with it and shut up about it."

Brain science, and biological anthropologist Dr. Helen Fisher, say something similar: "We look in the brain, at happy relationships, and try to see which parts of the brain become active in a really good relationship. And the main part is a brain region linked with what we call 'positive illusions,' the simple ability to overlook what you don't like about a human being and focus on what you do like." ∎

"I've always thought of my body as kind of the last frontier. My G-spot is a place no man has dared to go."

—JOAN RIVERS, COMEDIAN

"A woman has to take responsibility for her sexual satisfaction . . . Even the best lover, even one trained by me, can't bring her to orgasm unless she tells him what she needs."

—DR. RUTH WESTHEIMER, SEX THERAPIST

BACK TO BASICS

Does Size Really Matter?

One of the more compelling arguments to Pluto's reclassification to dwarf planet, of course, is that it's tiny compared to the eight planets. Among humans, size does not necessarily involve a demotion, though. On that point, sex therapist Dr. Ruth is very clear. "Neil, tell all of your viewers, size of that part of the male anatomy that we are talking about, the penis, size does not matter. A woman's vagina can accommodate all sizes. Now, if somebody has a miniscule one, that is a different story. I send them to a urologist."

Expanding Our Perspectives With Susan Sarandon

What's It Like to Be a Woman in a Man's Body?

Our scientific understanding of variations in sex and gender is still in its infancy. There is, though, increasing awareness that sex and gender "assignment" (given to a person by others, usually at birth, and usually through observation of genitalia) can differ from sex and gender "identity" (experienced by the person as an individual). Might there be physical causes for such differences?

"Freedom—it so opens up every-body's definition of who they are and what they can be . . . It opens up if you're not described by your genitals."

—SUSAN SARANDON, ACTRESS AND ACTIVIST

"One of the things I think makes a difference between male and female is testosterone," says advice columnist Dan Savage. "There's been some really interesting stuff written by people who were born into— you know, coercively assigned— female at birth, people who were born into women's bodies who were men, who then transitioned to male and took testosterone. And they have written about how their sexual thoughts, fantasies, everything, radically changed after testosterone."

Scientifically, transgender issues also appear to be independent of issues of sexual "orientation," which describes a person's romantic or sexual attraction to others. How will scientific understanding change the way our society deals with love and life? ∎

TOUR GUIDE

Do Kung Fu Nuns Use Nunchucks?

The Buddhist leader His Holiness the Gyal-wang Drukpa made a revolutionary decision to teach nuns martial arts. "Is that where we got nunchucks?" asks Jason Sudeikis, comedian.

Maybe not, but His Holiness did have excellent reasons for this new education: "One of the main targets . . . was gender equality . . . Kung fu can give the confidence and the defense. Not to harm others, of course—that is not my main aim—but to defend them . . . I could feel some resistance and some discomfort [from some men], but of course now they have to get used to it, because I will not give it up. Anyway, I'll keep on doing it, and I'll keep on dedicating my life to gender equality."

THINK ON THIS ▶ **Let's Do the Time Warp Again?**

"I do think that the difference between men and women is really gray, and so in terms of problem-solving, or imagining, or empathy, or any of those things, [it's good] to have both genders. And now we're getting genders that are transgenders, so the crayon box is even bigger and everyone can color outside the lines everywhere."
—Susan Sarandon, actress and costar in the gender-bending musical movie *The Rocky Horror Picture Show*

Boy? Girl? What's the Difference?

The biological differences between males and females are obvious. The complexity of human physiology and behavior, though, often blurs the distinctions. "I think men and women are like two feet," says Dr. Helen Fisher, biological anthropologist. "They need each other to get ahead. But for millions of years, they did different jobs, and that built, really, some differences in the male and female brain."

And the questions of gender difference are getting broader every day. Astrophysicist and transgender woman Dr. Rebecca Oppenheimer explains her experience: "I always had a sense of who I was. However, I am just one of many billions on this planet, and many, many others feel quite differently from me . . . Science actually may not be simply about classifications . . . With gender come deep questions about a person's most innate sense of who they are."

▶ IS GENDER PREJUDICE INNATE?

Scientific research shows that biases of any kind are almost always learned behaviors. Often they are learned very early in life and they can pervade every part of human perception and behavior without most people realizing it. "People are more in the grip of their biases than they want to admit," says author Malcolm Gladwell.

Decades ago, a major U.S. symphony orchestra tried to reduce sex discrimination by holding blind auditions. During the tryouts, the judges wouldn't be able to see the auditioning musicians. But the male judges still made decisions biased against female applicants. It turns out, they could hear their heels clicking on the floor! Today, U.S. orchestras are much more gender-balanced—and auditioners are asked to wear soft shoes. ■

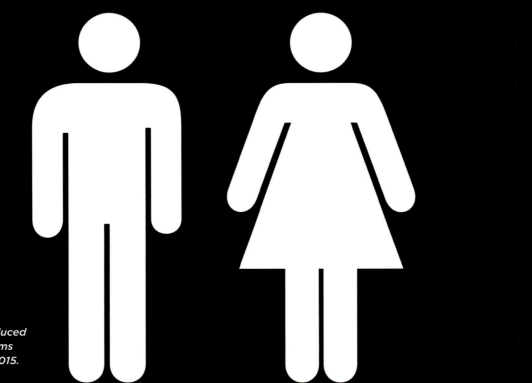

...t Obama introduced ...neutral bathrooms ...White House in 2015.

"The truffle is $1,500. The peach is $1. Which is better? The truffle is rarer, it costs more money, but is it more delicious than a perfectly ripe peach in season? Or a pear?"

—ANTHONY BOURDAIN, CHEF AND TV HOST

CHAPTER THREE

What's the Spice of Life?

Food! Glorious food! Thousands of years of growing, raising, and cooking food have created a legacy of deliciousness. The typical spice cabinet has dozens of flavorings, a typical kitchen a hundred ingredients. With practice, we all can make whatever we want to eat, exactly when we want it.

Indeed, we have gone past producing the optimal amount of food available for all, vaulted over the point of excess, and then kept going. For almost the entire history of humanity, the acquisition of nutrition has been a life-or-death struggle. Today, more and more of us have so much food that we're actually suffering the health consequences. In part, it's our ancient cravings for sugar, salt, and fat, combined with science funded by business interests. Happily, there's plenty we can do to fend off the bad effects of this gluttony in our daily lives while still eating well at home and as we travel.

So eat, drink, and be merry! Not too much, though—for tomorrow, we shall dine again.

Could the science of food lie in an array of spices and herbs?

Everything You Need to Know About Salt

Ancient Roman soldiers were paid with salt—hence the term "salary." Even though it's cheap today, salt has tremendous power. Its food-preserving and taste-enhancing properties have shaped our society—and our bodies, too.

|||||||

◄ WHAT MAKES GOURMET SALT COLORFUL?

"Until modern times, a goal of salt producers was to try to get all that stuff out and try to get it as white as possible . . . Every color is an impurity." —Mark Kurlansky, author of *Salt: A World History*

|||||||

▶ HOW MANY USES OF SALT ARE THERE?

"The U.S. is the largest producer and consumer of salt in the world . . . And guess what—only 8 percent of that is used for food . . . The industry claims 14,000 uses for salt." —Dr. Neil deGrasse Tyson, astro-salt-icist

|||||||
▲ HOW DOES NEIL KEEP HIS BROCCOLI SO GREEN?

"Put in a dash of salt while you're boiling vegetables and it keeps the color of your broccoli bright and green, rather than that faded green you get in canned vegetables."
—Dr. Neil deGrasse Tyson, astro-broccol-ist

|||||||
▲ WHAT DO SALT AND HEROIN HAVE IN COMMON?

"The biological hardwiring in our brain for salt follows the same pathways and nerve cells as some addictive drugs." —Neil

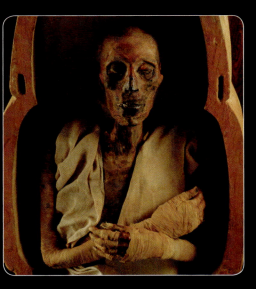

|||||||
▲ CAN SALT REALLY KILL YOU?

Ingesting too much salt at once can cause seizures and death. That's a rare concern, though—you would have to eat several cups for that to happen. A little too much salt each day—say, 0.03 ounce, or about 1,000 milligrams—is the more insidious danger.

FOOD FOR THOUGHT
Are You Worth Your Salt?

The ancient consideration of salt as a high-value commodity shows through in our language today, including in the phrase "to be worth your salt." "Basically the entire food trade depended on salt," says Mark Kurlansky, author of *Salt: A World History*. "And in preindustrial society, that was a very large part of trade. So it's not an exaggeration to say that without salt, you couldn't have an international economy."

And it isn't just in western European history that salt has a value going back centuries. "In the southwestern pueblos, it was very much of a commodity that people traded back and forth, and that goes back 3,000 to 4,000 years . . . Hopis still have a salt pilgrimage that they go on . . . to this day," says anthropologist Dr. Peter Whiteley. ■

Wit and Wisdom About Wine

What's Up With Yeast—Is It Really Alive?

Yeasts are single-celled fungal organisms that, like us, eat sugar and release carbon dioxide. They also produce alcohol as a by-product. So if we want gas or booze in our food, yeast is our go-to ingredient. "If you let nature do its thing, there are thousands, if not millions, of types of yeast, and just like people—we all operate differently, look differently, smell differently—so do yeast," explains Jennifer Simonetti-Bryan, Jedi Wine Master. "So when you have that complexity of all of them working at the same time, they all produce something different, and so there's layers of flavors, which we call in wine 'complexity.'"

"Then the yeast dies in its own excrement," adds Neil. "And that's Winemaking 101."

Under normal conditions, many kinds of yeast are harmless to humans. So we owe to yeast all our wine, beer, and whiskey, not to mention bread, kefir, kimchi, miso, and lots more. ∎

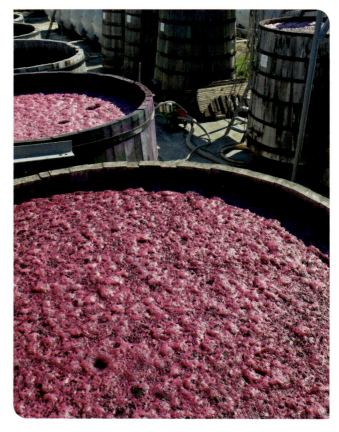

Oak barrels flavor fermenting wine.

"Vanilla is genius. People say vanilla is boring. I go, 'Are you kidding?' Vanilla is perfect. It's a perfect flavor . . . If you want your mint chocolate-chip, if you want your peppermint stick, knock yourselves out. 'Cause it's more [vanilla] for me."

—BILL NYE THE VANILLA GUY

THINK ON THIS ▶ Tasty or Deadly?

"All-natural ingredients" doesn't always mean all-healthy foods. Plenty of plants produce toxins that can kill us quite quickly—but that's where a lot of our medicines come from. And our spices, too. Nicotine, caffeine, cinnamon, and vanilla are all potent insecticides. "Plants can't run away, so they're producing things that can protect themselves from insects or herbivores, and we . . . have figured out how to leverage these things for our benefit," says Dr. Mark Siddall, biologist.

Wit and Wisdom About Wine

What Is Terroir, and Why Should I Care?

Most people agree that the grape-growing environment matters. "There's a French term called terroir, referring to everything around the vine that makes those grapes from that vine taste the way that it does," says Jedi Wine Master Jennifer Simonetti-Bryan. "People have different definitions, but it's that 'je ne sais quoi' that makes it taste the way that it does."

Really, though, how much does it matter? Once the bottle is opened, so much can be done to the wine that will ultimately affect its flavor even more. "You know how people will decant wine so that it can breathe?" asks Neil. "[Nathan Myhrvold, founder of the Cooking Lab] takes wine and he puts it in the blender. He gave it to expert tasters and it improved every bottle that he did it to. And then he told them that he put it in a blender, and all of a sudden they didn't like the wine." ■

A vineyard in Tuscany, Italy

NEIL TWEETS

What Are Neil's Favorite Mixed Drinks?

"I like foofy drinks. I go to a bar, guys are ordering whiskey, and I order something with a pineapple wedge and an umbrella in it. I love it. I am comfortable enough with my masculinity, or rather I am in touch with my feminine side enough, that I have no issues drinking an umbrella drink in a bar. But I also lean towards the creamier drinks, so that means over the holidays I'm an eggnog guy from way back. Spiked eggnog . . . And when I'm ambitious, my wife and I will make it ourselves. It's effort, but it's worth it." —Dr. Neil deGrasse Tyson, astro-foofy-cist

THINK ON THIS ▶ Should You Open Champagne With a Saber?

Sabrage is the art of slicing open a champagne bottle at its neck with a sword. It's an impressive ceremony, but is it worth the risk? Sometimes . . . "I actually had to use this method [once]. It was an enormous bottle, an imperial [the equivalent of eight bottles]. Half of the cork was still inside the bottle—it was a dangerous situation . . . I happened in my office to have the saber, and I saved the day," says Jennifer Simonetti-Bryan, Jedi Wine Master.

Are Yummy and Yuck Acquired Tastes?

U.K. consumers tasted oddball flavors of Walkers chips and voted for their favorite.

Chef and TV host Anthony Bourdain shares how he enjoys the variety of global cuisine: "I got over using words like 'bizarre' a long time ago when looking at how other people eat around the world. There are whole spectrums of flavors that other countries, other cultures, take for granted and require in their diet. In the Philippines, there's a whole bitter component that we are almost instinctively not happy with. They will introduce bile into dishes to give it that welcome bitter note.

"Cultures like Scandinavian cultures, where there's a very limited spectrum of flavors—not a lot of spices traditionally, a lot of fresh fish, fresh fish, frozen fish, more fresh fish, maybe some preserved fish—as well as South Pacific cultures, where it's all sort of sweet fresh fish, not a lot of salty, savory. There's a tradition of rotting things, like fermenting fish, getting it really offensively funky by our standards, just 'cause, I think, out of boredom. And it's worth noting also that we, Western societies anyway, used to do that. Roman times, the condiment of choice was something called garam, that was essentially rotten fish guts, and rotten fish sauce; this was the salt, the principal seasoning ingredient all across Europe. So even our own tastes have changed." ∎

DID YOU KNOW

In 1985 a medical research team proved that Coca-Cola—especially Diet Coke—was an effective spermicide. The team did not find, though, that Coca-Cola was an effective method of birth control.

THINK ON THIS ▶ Can Candy Make You Smarter?

DR. MAYIM BIALIK: The issue there is motivation, and not necessarily a skill set and a cognitive ability or a technical ability. The fact is, yes, candy makes everything better, no matter what you're trying to learn, because it's a very strong motivator.

DR. HEATHER BERLIN: It might not make you better at math, but it might make you study for longer.

A Seat at the Table With Anthony Bourdain (Part 1)

Too Little or Too Much?

B illions of people worldwide don't get enough to eat, and yet Americans are still getting fat. And much of the rest of the world, alas, is trending our way. "What's happening in the developing countries now is that as everybody gets a little money, they start eating more, and they start eating like we do, and they put on weight and develop type 2 diabetes, and there it goes," says Dr. Marion Nestle, author of *Food Politics*. "It even has a name. It's called the 'nutrition transition.'"

▶ HOW DO I KNOW WHEN TO STOP?

"We have about a hundred physiological factors that encourage us to eat more," explains Dr. Nestle. "We're not very well tuned to the environment that we're in, and our physiology is much better at saying, 'Eat, eat, eat, eat. You're hungry—better get the glucose to the brain quick,' and much, much less effective at telling us when to stop."

▶ WHY ARE VEGANS BETTER OFF IN NONINDUSTRIALIZED COUNTRIES?

"One of my favorite statistics is that apparently vegans in nonindustrialized cultures seem to do much better than vegans in industrialized cultures," says chef and TV host Anthony Bourdain. "Apparently, the insect parts and carcasses in rice are much higher in nonindustrialized cultures, so basically they're getting a lot more animal protein. They're very high in protein—bugs—by the way."

▶ HOW MUCH DOES OBESITY COST THE U.S.?

Dr. Nestle offers one figure: "There have been estimates—I don't know how good they are—that overweight costs America $190 billion a year."

"You can go to Mars twice for that," adds Neil. ∎

Is the Food Industry Killing Us?

Since time immemorial, people have sold stuff to other people, whether or not it's beneficial or necessary. You see it every day in the supermarket aisles, says chef and TV host Anthony Bourdain: "With the good comes the bad, and the bad might be that it is in the financial interest of some very large, powerful companies that you continue to eat badly, and too much. And they're going to continue to spend a lot of money, as any company will do, to make you continue to buy their products. And a lot of these products are not ideal staples of any diet."

Sometimes the public gets a glimpse of what's really going on. Consider the recent dustup over "pink slime." Bourdain explains: "Pink slime is not an ingredient, according to the rules. It is a process that allowed ground-beef manufacturers to essentially buy the outer cuts of beef that would otherwise previously have had to be discarded . . . [because they can] contain *E. coli* . . . By introducing, as I understand it, an ammonia vapor—basically steaming this stuff—and whipping it into a mulch-like paste with bits of extruded fat—mixing it into this slime and processing it with ammonia—they were able to bring the likelihood of *E. coli* down." ■

The notorious "pink slime" beef

BACK TO BASICS

What Is the Essence of a Peach?

In the secret vaults of food-company laboratories, scientists have for years sought to isolate and re-create the tastes, smells, and textures of the foods we know. Artificial food products like Velveeta, Jell-O, and Tang permeate modern food culture. What about the highbrow version of this edible engineering— the molecular food movement?

"It's treating ingredients in new ways," says chef and TV host Anthony Bourdain. "It's manipulating preexisting ingredients into unusual forms . . . to trick the mind into eating a strawberry that doesn't look like a strawberry, an apple that looks and feels like caviar in the mouth . . . It's not chemistry class, but it certainly does look like a laboratory."

"The fact that McDonald's and other retail outlets are saying, 'We're not using it any more'—it's not like they're nice guys. They're looking pretty far into the future and seeing, 'This is going to come back and bite us.'"

—ANTHONY BOURDAIN, CHEF AND TV HOST, ON RESTAURANTS' USE OF "PINK SLIME"

A Seat at the Table With Anthony Bourdain [Part 2]

How Do I Avoid Getting Sick While Traveling?

t's tempting to try every possible exotic food when you're traveling. Street food vendors are often nearly irresistible. In almost every case, though, you'll be better off eating properly washed and fully cooked things. Even if that may kill the buzz in trying the most enticing local delicacies, you'll generally be healthier if you kill the harmful microbes in your food.

Here's some more stomach-saving advice from globe-hopping chef Anthony Bourdain, who's been around the block enough times to know from experience what not to do:

▶ **1.** "Exercising reasonable caution, the same way you would if you travel around rural America, is a useful thing to do wherever you go."

▶ **2.** "Always ask yourself, is this how your average person eats. Is the place busy?"

▶ **3.** "If you're aware that avian flu has become a concern in the area, undercooked poultry is probably not going to be a good idea . . . You have to think about those things. If there's mad cow [disease] in the area, maybe calf's brains at a dodgy pub would not be your first option."

▶ **4.** "Just common sense. If they're not drinking the water in Russia from the tap, maybe you shouldn't either." ■

> "Cooking does wonders for food safety . . . You will be so much better off eating cooked food in places where the water's dirty."
>
> —DR. MARION NESTLE, AUTHOR OF *FOOD POLITICS*

Chef and TV host Anthony Bourdain has put his stomach to the test.

> "I think [people eat blowfish sushi] because of the risk of dying . . . Tetrodotoxin is the poison . . . What's amazing is that it does not affect your heart, so you get to stay alive while you're dying."
>
> —DR. MARK SIDDALL, BIOLOGIST

THINK ON THIS ▶ **Is Drinking Blood From a Beating Cobra Heart Safe?**

Who drinks fresh cobra blood? Anthony Bourdain does—or did. And lived to tell the tale: "[There was a time] when I would eat as far out of my comfort zone as a daredevil, just so that I could tell friends I drank live cobra blood. I don't do that anymore, and I guess I would advise people against [it]." A cobra's blood has no venom in it, as long as you stay clear of the fangs.

"I was either going to go to science, engineering, all that, or an art school . . . To me, there seemed to be no difference in the kind of creativity and imagination that goes into either of those areas. I thought, They're equivalent."

—DAVID BYRNE, MUSICIAN

CHAPTER FOUR

Where Does Creativity Come From?

n 1930 Albert Einstein wrote that the impulse to wonder about the unknown "is the source of all true art and science." Where does that impulse reside? Scientists haven't figured that out yet. Maybe it's locked in the myriad patterns of synapses—the spaces and connections between brain cells. Maybe it's something external to us—a divine spark or an emergent consciousness. Or maybe it's hiding in our dreams, waiting for us to fall asleep so it can be properly placed in the circular files of the mind. How funny would that be!

Einstein, perhaps the most creative man of the 20th century, was hardly a "normal" person—and the world was better off for it. Then again, what does it mean to be a normal human being? If we can be sufficiently self-aware to cast aside our preconceived notions of the characteristics that define typical people, entire new realms of discovery and creativity may spring forth from new directions. That sure sounds good for us all—and it sure feels good, too.

Children's creativity is incited by their untamed imagination.

$E=MC^2$

x

4cm

5cm

StarTalk Live!: Big Brains at BAM (Part 3)

What Is the Neural Basis of Creativity?

B iologically, creativity is essential for human survival. If we run into a life-threatening situation that we've never experienced or heard about before, we need to be able to improvise a solution on the fly. Creativity, in our fight or flight mode, has enabled us to make it this far as a species.

So our brains must have developed some kind of wiring that specifically supports creative activity. How exactly might that work? Let Dr. Heather Berlin, neuroscientist, explain: "What we find is that, whether it's jazz improv or comedy improv, there's a certain neural signature involved when people are improvising. So you can put people in a scanner, even freestyle rappers, and you give them a memorized rap, and then they can freestyle. Or you give a musician a memorized piece or they improv. And when they're improvising . . . a part of the medial prefrontal cortex becomes extra-activated, and that has to do with internally generated ideas. And the dorsal lateral part of the prefrontal cortex becomes deactivated, and that has to do with sort of self-awareness and monitoring your behavior. So you almost go into this free-flowing state. If you become too aware, you mess up, you're not good at improvising. You have to kind of lose yourself, so to speak." ◼

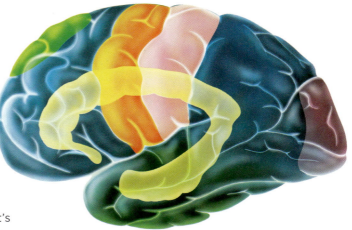

The brain turns on self-awareness to generate pure improvisation.

"I wrote a paper on the physiology of laughter in college. It was all wrong, but I wrote the [bleep] out of it."

—EUGENE MIRMAN, COMEDIAN

THINK ON THIS ▶ **Does Talent Come Naturally?**

True genius unquestionably exists. Some people can just do things the rest of us can't, no matter how hard we try. Many people think, though, that natural talent isn't enough—or, maybe, hardly even matters. Time, practice, and experience, they argue, are the key. "What we associate with Mozart's genius is stuff that he produces 14, 15 years into his life as a composer, which is a fascinating and sobering thought," says author Malcolm Gladwell.

But Will I Be Able to Play the Piano?

Does it matter from where inspiration strikes? Neurologist and author Dr. Oliver Sacks told the story of an acquaintance who suddenly gained a new personality trait after being hit by lightning and almost dying: "About three weeks after this, he had a strange emotional and musical change. This man, who had never been interested in music, developed a sudden passion for classical music. First to hear it, and then to play it. And then he wanted to compose it. And this also went with a mystical feeling. He felt that God had sent the thunderbolt but had also arranged for him to be resuscitated, and that he now had a mission to bring music to the world.

"This man put a supernatural explanation on this. However, he is not ignorant scientifically—in fact, he has a Ph.D. in neuroscience as well. As a neurologist, it was up to me to put things in more neurological terms, but without in any way upsetting him, devaluing his experience. And I said, 'You know, I'm sure this is what you experienced and what you believe, but will you allow that something might have happened inside you? For example, might supernatural intervention make use of existing neurological structures?' And he said, 'Yeah, okay.' "

"It's a mysterious mechanism, acting and theater and storytelling . . . It is a mystery to actors as well."
—ALAN RICKMAN, ACTOR

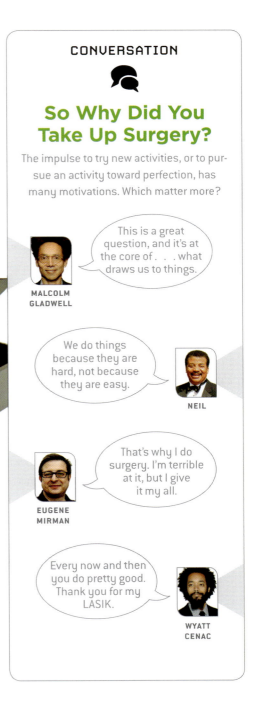

DID YOU KNOW

A typical human has about 90 billion brain cells, 20 to 25 percent of which are in the cerebral cortex, which governs language and consciousness.

CONVERSATION

So Why Did You Take Up Surgery?

The impulse to try new activities, or to pursue an activity toward perfection, has many motivations. Which matter more?

This is a great question, and it's at the core of . . . what draws us to things.
MALCOLM GLADWELL

We do things because they are hard, not because they are easy.
NEIL

That's why I do surgery. I'm terrible at it, but I give it my all.
EUGENE MIRMAN

Every now and then you do pretty good. Thank you for my LASIK.
WYATT CENAC

StarTalk Live!: Big Brains at BAM [Part 1]

How Do I Know I Have Consciousness?

René Descartes, the 17th-century rationalist philosopher, opined that the thought "I exist" must be true because it's impossible to refute: *Cogito, ergo sum*—I think, therefore I am. Twenty-first-century scientific researchers want to go deeper than that as they probe the connection between awareness and consciousness. "We can define consciousness very simply as first-person subjective experience," explains neuroscientist Dr. Heather Berlin. "So you only are aware that you have it. I don't know what your consciousness is like; I only know what it is from [my own experience] internally. How is it tied to the brain? We're still trying to figure that out. Now, that's different from self-awareness. So you can be conscious without being self-aware . . . For example, babies. They can be conscious, meaning they have raw sensations—like seeing the color red or feeling something soft or smelling a rose—without being aware of oneself, or having sort of metacognition, like thoughts about other thoughts or 'I am the one having these thoughts' . . . There are certain dissociative disorders where people lose their sense of self, but they're still conscious."

Eugene Mirman puts it another way: "So you're saying a baby could hear Bruce Springsteen but not know why it's having so much fun." ■

DRINK OF THE EVENING

The Brain Freeze

This cool cocktail was concocted by Dr. Neil deGrasse Tyson and Brian Ponce, the bartender at the Brooklyn Academy of Music (BAM).

2 oz. vodka
A splash of triple sec
A splash of lime juice
A splash of cranberry juice
A splash of pineapple juice
Ice

Fill ⅔ of a tumbler with ice; add all other ingredients. Shake, then strain into a martini glass. Garnish with a slice of lime, as an homage to the fact that there are no green stars.

THINK ON THIS ▸ What Causes Déjà Vu?

"The notion is that there are many ways to get to one destination, and sometimes pathways get triggered that have this sense of familiarity because you've literally activated a pathway that is redundant . . . [which also has] that sense of 'I've done this before.' And your brain thinks it has." —Dr. Mayim Bialik, neuroscientist and actress

Every year, nearly 60 million Americans are affected by sleep disorders.

Why Do I Have to Sleep? Why Can't I Stay Awake Forever?

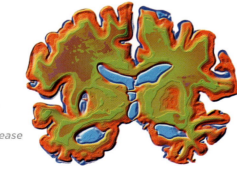

A cross section of a brain with Alzheimer's disease shows severe degeneration.

Really? Eight hours? "Why the hell do we have to sleep? What a waste of time that is," complains Dr. Neil deGrasse Tyson, astro-snooze-icist. "If an alien came to Earth and you're having a great conversation and you have to say, 'Excuse me, I have to lay semi-comatose for the next eight hours; I'll get back to you,' they'd wonder what's wrong with you."

Scientists are fascinated by the question "Why sleep?" as well. Neurologist Dr. Heather Berlin answers: "The latest theory that most of the neuroscientific evidence is pointing towards is that it is a way for the brain to sort of clean itself out. During the day you're taking in all of this stimulation and it's almost impossible for your brain to kind of integrate all of it every day—it would be cluttered. So at night what happens is there's a sort of pruning. It solidifies the information that it wants to keep. They've looked at people when you deprive them of sleep and it causes all sorts of problems. It can relate to things like Alzheimer's." ■

Why Does Mental Illness Persist?

The human brain is so complicated and interconnected that when something goes wrong, it's very, very hard to isolate the exact cause—and even harder to fix.

Part of the problem in treating mental illness is recognizing it. Mentally speaking, what is normal behavior and what is illness? "There is no such thing as normal," says neuroscientist Dr. Heather Berlin. "But studies do show that [of all diseases] the thing that causes people the most distress is mental illness. Because they don't die from it, but they have to suffer."

"Often, they have children and pass on genes that may be predisposing to future mental illnesses. It's a really sad discussion."

—DR. MAYIM BIALIK, NEUROSCIENTIST AND ACTRESS

Society, correctly or incorrectly, often associates abnormality or instability with tenacity and genius. And it's certainly not in the best interest of society to quell creativity or diversity of thought and action. "Certain things do persist," says Dr. Berlin. "Like, for example, being a sensation seeker, being really impulsive: That's a quality that got people out there, and discovered America, right?"

Whatever the cause, the consequences of not properly diagnosing mental illness can be tragic. ■

The social stigma of mental illness is a barrier to treatment.

Funny Ha-Ha or Funny Strange?

Neuroscientist Mayim Bialik may be best known for her role as Amy in the television sitcom *The Big Bang Theory,* known for the quirky humor generated by its even quirkier characters. So what would she say about them?

"All of our characters are, in theory, on the neuropsychiatric spectrum . . . I think what's interesting and kind of sweet—and what I think should not be lost on people—is that we don't pathologize our characters. We don't talk about medicating them or even really changing them . . . This is a group of people who likely were teased, mocked, told that they will never be appreciated or loved . . . who have successful careers, active social lives that involve things like Dungeons & Dragons and video games—but they also have relationships, and that's a fulfilling and satisfying life." ■

Neil clicks with the characters of The Big Bang Theory.

BACK TO BASICS

What Is Savant Syndrome?

Savant syndrome is a psychological condition in which you may have a mental handicap in one or more significant ways and yet you can perform some cognitive functions at almost superhuman levels. The most common kind of savant you hear about is calendrical: Name any date in the past, and they can tell you what day of the week it was. Mathematical savants can make certain calculations in their head rapidly and flawlessly. Musical savants might be able to play Beethoven sonatas perfectly by ear. Although about half of the time, savant syndrome appears in people who are on the autistic spectrum, scientists still have almost no idea how or why it appears.

THINK ON THIS ▶ What Is Visual Thinking?

Dr. Temple Grandin, a renowned animal scientist and autism activist, has been formally diagnosed with autism spectrum disorder. She explains that her type of photo-realistic thinking is one of the ways that the autistic brain can differ from the typical brain. It allowed her to "get into the head of cattle," noticing how they feared shadows and reflections, which everybody else had missed. "My mind thinks kind of like Google for images," she says.

Is There a Science to Comedy?

This is a question that Neil has asked nearly every comedian who has been on *StarTalk*. While they might not agree on a single formula, each of them takes the craft very seriously and applies a version of the scientific method. Except for Joan Rivers . . . You can read what she says for yourself.

"Comedy is all about math, too. It's all about having the correct amount of words . . . There's the correct amount of words to make a joke perfect. And once it's there, it's done. You're done with that joke; move on to the next one."

LARRY WILMORE

"You're going to find structure and science is more in the sitcom realm, because it's setup, delivery, punch, setup, delivery, punch. It's formulaic. In sketch comedy, it's less formulaic; it's more absurdist. You're never really sure of an ending, so you're not so conscious of bringing the scene to a peak, and then . . . an anticlimax, and then finishing it on a climax . . . Sketch comedy is much more ephemeral . . . Of course the purest form is [improv at] Second City. Second City is mercury rolling across the table and splitting up into little balls. Sitcom or film is more of a structured, molecular picture, where you're actually designing and confining things."

DAN AYKROYD

"There is a science to a joke. And one of the things that you'll find . . . [is that] the longer the setup, the harder it is for you to get the payoff. So the shorter the setup, the bigger the payoff. That is, if it's funny."

CHUCK NICE

"Al Jean, who's been running [The Simpsons] for years . . . comedy is like math to him. The script becomes this equation that he's figured out, that pays off in the right places."

HANK AZARIA

"I think that stand-up is very much the scientific method . . . You go onstage; you try something. If it works, you keep it; if it fails, you switch it out."

EUGENE MIRMAN

"For me, as an actor and as a scientist, what I'm aware of is, we're constantly scanning and tracking . . . It's a very complicated process by which I need to make you feel something, I need to make you believe something, and I need every single person to feel that. So, if you're a stand-up, you're working with a room of people. When I work in front of a live audience, I need everyone to feel something. And it's a really complicated interaction."

MAYIM BIALIK

"Anyone who thinks they've gotten it down to ones and zeros has lost what makes comedy great. With that said . . . when you rewrite sketches, just by trial and error you can sort of say, 'Hey, as somebody who's seen 500 sketches, I will just tell you that [works]' . . . There's a little bit of experimentation and finding results and sort of tracking them, but they're not absolute rules."

SETH MEYERS

"Comedy is not controllable, because you can think something is funny and nobody else does . . . But geometry you can control: This to that equals this . . . You can't change it, and I can't change it, and that's it . . . Comedy is not geometry. It is not a science. There's no such thing as the science of comedy. And people that try to teach it, I feel are so cruel."

JOAN RIVERS

LAUGH OUT LOUD ▶ With Jon Stewart

Neil asked actor and comedian Jon Stewart about his favorite element on the periodic table. His answer: "Oh, I'm a huge carbon guy. I enjoy the molecular slut of the table of elements." Why call it that? Easy. Jon says: "It will bond with anything."

Emoticons are a shortcut for expressing feelings.

"What's interesting when you study the way faces express emotion across cultures, there's extraordinary similarity. An angry person in one culture looks like an angry person in another culture."

—DR. NEIL DEGRASSE TYSON, ASTRO-PISSED-ICIST

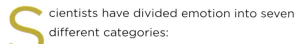

OK, So Tell Me: How Did That Make You Feel?

Scientists have divided emotion into seven different categories:

▶ **HAPPINESS** Scientists consider joy the most basic component of happiness; facial expressions of joy can be recognized across every social and cultural group.

▶ **SADNESS** Grief—the mental suffering due to loss or pain—underpins sadness, which can range from disappointment to heartache to despair.

▶ **ANGER** This feeling of dissatisfaction comes usually when some sort of wrongdoing has been perceived.

▶ **SURPRISE** The fundamental component of surprise is the suddenness of the emotion.

▶ **FEAR** The "fight or flight" emotion causes a massive release of adrenaline, instantly raising heart rate, blood pressure, strength, speed, and sensory acuity.

▶ **DISGUST** Eyes crinkle, noses wrinkle, the upper lip pulls up, and the lower lip loosens when we say, "Eww!"

▶ **CONTEMPT** Likely the combination of anger and disgust, contempt is directed toward lower-status entities. ▪

THINK ON THIS ▶ How Far Can a Good Joke Take You?

Thanks to the speed of radio waves and the vast distances to other stars, anything broadcast today moves outward from Earth at the rate of nearly six trillion miles per year. "If you parked next to Arcturus tonight, which is up in our sky, you would be hearing radio broadcasts or TV from 40 years ago," says Carter Emmart, astronomer and artist. So you better make sure your jokes are still worth listening to a long time from now.

Shower songsters exploit the prime acoustics.

The Science of Music With Josh Groban

Why Do We All Sound Good Singing in the Shower?

"There's no half singing in the shower. You're either an opera diva or a rock star. There's no James Taylor in the shower . . . It's all or nothing in the shower."

—JOSH GROBAN, MUSICIAN

The acoustical properties of a shower stall produce a feedback loop for any singer. It's like having six echoes coming back at you with every note you sing. "It's a pretty epic reverb in the shower," says musician Josh Groban. "That's like a karaoke reverb in there. No one sounds bad in the shower."

There's more to it, though. The sound of the water spray creates a white-noise screen that masks out-of-tune notes and less pleasant frequencies, so only the singer's prettiest primary tones come through. (Listen to a school chorus through the door of the music room for a similar effect.)

Here's another factor to consider—the psychological one. With all the noise in the shower, you can't hear the outside world, and it seems like the outside world can't hear you either. That can remove the singer's inhibitions. Voice teachers will attest that everyone sounds better when singing more confidently, when being self-conscious isn't an issue. So that means you're not the only one who thinks you sound better in the shower. ■

"I think the greatest art allows you to walk up to it and say, 'That means something to me' regardless of what the artist thought or felt."

—DR. NEIL DEGRASSE TYSON, ASTRO-ARTIST-IST

The Science of Creativity With David Byrne

Does Science Inspire Art?

Science is a major human endeavor by which we explore the unknown, and it always has and always will inspire art. The connection goes both ways: How many of today's inventions were first conceived in a science-fiction book, movie, or television show?

Here's a variation on the question: Can we apply purely scientific principles to a "soulless" system and produce art? For example, can we program a computer to compose music? Electronic composer Dr. David Cope thinks so: "The first [computer] program that brought a lot of notoriety was written in about 1980. It's a program called Experiments in Musical Intelligence, and it . . . takes a database of music by typically a classical composer (because they're dead and they can't sue me for imitating their style) . . . and analyzes that music, and fundamentally tries to create a new composition in the style of that music." ■

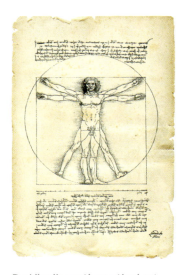
Da Vinci's mathematical art

TOUR GUIDE

Can Science and Art Together Achieve Greater Innovations?

"Whenever someone has an understanding of what people really like to do, what they really want to do, they can make good technologies."—Clive Thompson, author of *Smarter Than You Think*

In a college commencement speech, Apple co-founder Steve Jobs credited a college calligraphy class for inspiring his revolutionary work on the first Macintosh computers.

"I still kind of think like an artist," says Biz Stone, co-founder of Twitter. "When I build systems, I think of how it will make people feel, and how they will make others feel."

THINK ON THIS ▶ Is Science Your Muse?

"If I use the word 'nova,' I'll be speaking about a star—unlike the average rapper, who will probably be talking about a car," GZA, co-founder of Wu-Tang Clan, tells us. Of course, both nonscientific and scientific views of the cosmos can inspire artistic expression: "The wonder for the universe happens to a person only when they get out on a clear night . . . and they're starstruck . . . and that's the moment I want to try to communicate," says musician David Crosby.

Where Does Art Live?

Artistic expression may not live in any particular realm of the brain. The human creativity that drives art may indeed be the ability to reach across or between realms to get to that otherwise intangible destination. "We are part of something that is so enormously big," says artist Peter Max. "There's science, there's the mystery of the science, because there is so much more that we don't know than we do know . . . Through meditation I was able to get very, very peaceful and quiet, but through looking at the universe I would get excited, and so in between the peacefulness and the universe is where the art world lived."

What about computers that produce art? Neil isn't so sure we've reached that state of technology yet:

"You need a computer that can have a broken heart."
—CHUCK NICE, COMEDIAN

"Personally, I think that the computer doesn't yet know how to feel emotion. And what is art without emotion? . . . If it's a computer just punching out notes according to some algorithm, I don't know if it can reach those same heights."

When humans experience beauty, said the philosopher Immanuel Kant, it is not for a practical purpose; rather it is an opportunity for us to see the highest and best things possible in the universe, for their own sake and not for our own. So to Kant, art bridges the gap between the mundane and the sublime, living in a dynamic space between the two. Is this also the space between computers and emotions—or between peacefulness and the universe? ■

Oil paints transcend their original form in the hands of an artist.

"We don't see any reason to make that line between kids and adults and play. [Kids] are understanding their world through these adventures that they're having, the little experiments."

—JAMIE HYNEMAN, COHOST OF *MYTHBUSTERS*

CHAPTER FIVE

Shall We Play a Game?

Let's pretend. What we do right now won't be for keeps—we'll take a break from the real world, and we won't have to deal with the consequences of our actions. How 'bout it?

I'm sorry. I'm afraid we can't do that. Everything's real these days, even the virtual world; we can't tell the difference between real and imaginary any longer. Technology is confusing the artificial with the natural. Our favorite sports are getting too vicious to play. And even our minds and our senses play tricks with us as we use psychotropic agents to find new and unusual ways— maybe dangerous ways, too—of perceiving our surroundings.

Well, in that case, let's play in real life. We'll have fun while we figure out how the universe works. We'll use video games to improve our social skills. We'll observe science in action as tilted cars zip by. We'll ride on rising and falling waves of metal. And we'll watch how our entertainment media can make the world a better place. Let's go!

Video games offer endless virtual entertainment.

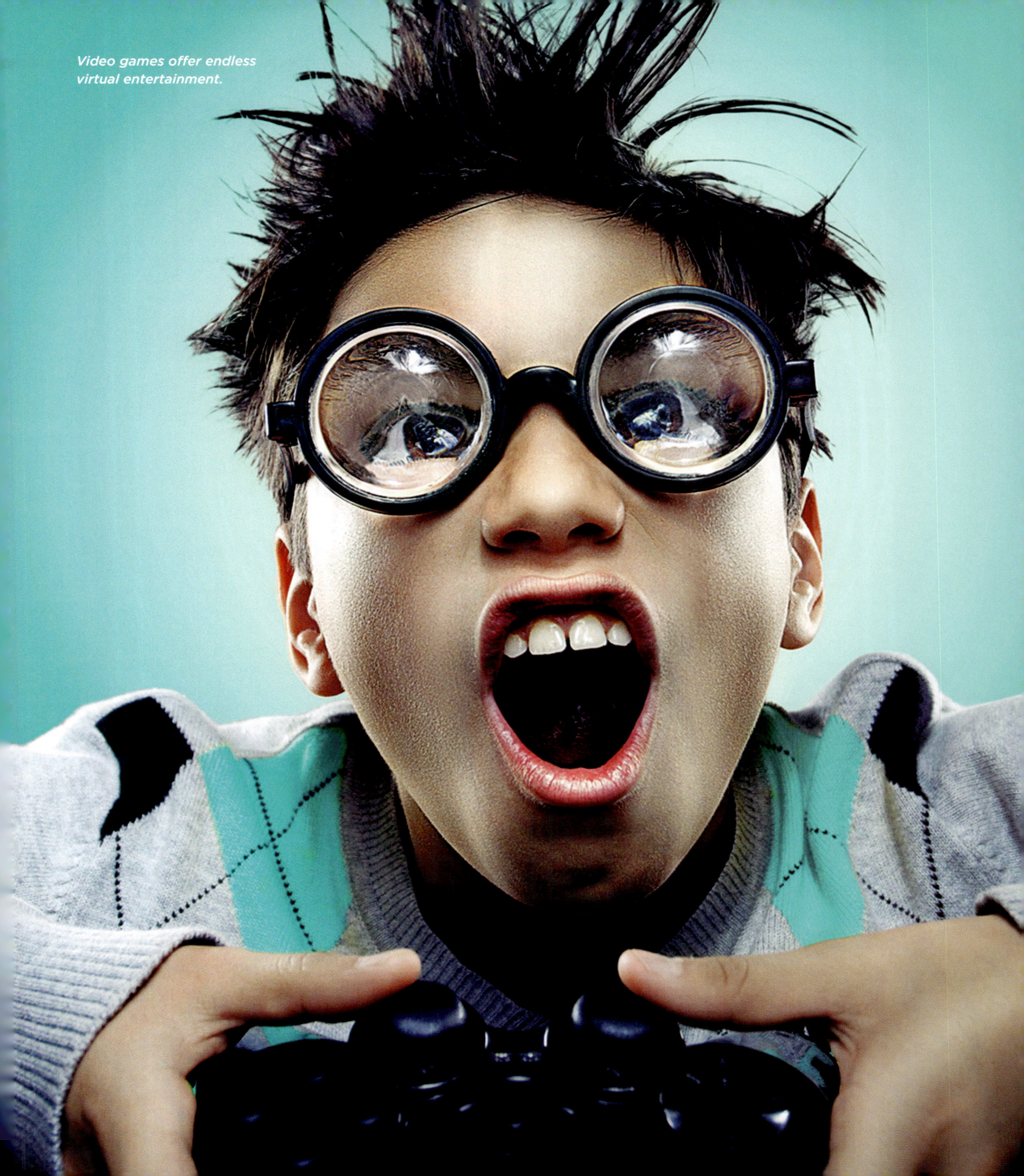

Cyber games may simulate realistic violence.

Can Video Games Create Violent People?

Tablets and smartphones make gaming mobile.

Since time immemorial, games and sports have been designed to train children for real life. It's probably an extension of the animal kingdom, where tiger cubs and wolf pups wrestle in training for real hunting. Are the games of today really so different from those of yesteryear? "It's interesting that kids, in the absence of video games, used to play Cowboys and Indians, or Cops and Robbers," says Will Wright, designer of *The Sims.*

In our time, parents and policymakers have worried about one bad influence after another: comic books, television shows, rap music, football, and, of course, video games. There is no scientific consensus about whether video games cause their players to act out in real life what they do on the screen. Prolonged exposure to certain fictional behaviors can either desensitize or hypersensitize people—they may become either less concerned about real-life violence or overly concerned and fearful of it. As with most things, moderation and reasonable limits seem to be the way to go. ∎

What's Real and What's Virtual?

In the early days of video games, even the simplest simulations were virtual versions of reality that pulled kids in. "I must have been 9 or 10," Elon Musk says of his childhood Commodore VIC-20 home computer. "You construct a little universe . . . You can actually make things happen. You can type these commands, and then something happens on the screen. That's pretty amazing."

> *"Even in Iraq, after they've gone and done their patrol, they come back to their tent and play* Counter-Strike *on their Xbox."*
>
> —WILL WRIGHT, DESIGNER, *THE SIMS,* ON SOLDIERS PLAYING WAR-BASED VIDEO GAMES

One of the first smash arcade video games was *Asteroids* (1979), where the player blasted space rocks and flying saucers to survive. "I spent way too much of my childhood playing that game," remembers astrophysicist Dr. Amy Mainzer. "And so when I became an asteroid scientist, it turns out the game is actually not that bad. If you hit an asteroid that is big in the game, it breaks into a lot of little pieces. And sure enough, that is exactly what happens."

Today we have remote-controlled war machines in the hands of young military personnel. "The generation of soldiers that are going into the army and flying these drones grew up with video games," says *The Sims* designer Will Wright. "They're still playing these games, and I know a lot of them are using them for teamwork exercises." ∎

TOUR GUIDE

Did a Video Game Actually Pass the Turing Test?

In 1950, computer scientist Alan Turing suggested a test for artificial intelligence: AI succeeded, or passed the test, when a human couldn't tell if answers to questions were coming from a man or machine. "There was something about 10 years ago called *Starship Titanic* by Douglas Adams," says Jeffrey Ryan, video game expert. "It was designed to have a free-flowing [text] conversation with you . . . In fact, they had Monty Python guys come in to work on it, to make sure the humor [worked]. It passed the Turing test for a little bit."

THINK ON THIS ▶ **Can Playing Video Games Increase Your EQ?**

"The most educational video game that ever was might have been something like *The Sims*," says video game expert Jeffrey Ryan. "Because think about how often you talk to other people in life and actually have the same sort of interactions." Training you to be socially aware might be upping your "emotional quotient," or EQ.

"We say this at the end of our touring show: that someone once said that the phrase that typifies real discovery is not 'Eureka!' but 'Oh, that's funny.'"

—ADAM SAVAGE, COHOST OF *MYTHBUSTERS*

MythBusters (Part 1)

Is Science Just Playing Without the Rules?

Scientists do follow some rules they have created and agreed on—the need to test their hypotheses, for example, with experiments and observations. Overall, though, science is trying to learn the rules of nature—and to have fun doing it. As the hosts of TV's *MythBusters* assert, playing around and trying out new ideas captures the essence of scientific discovery.

"It was somewhere in the second season we realized, wow, actually the structure of the show works best when we are having the most fun," says cohost Adam Savage.

Cohost Jamie Hyneman agrees: "You look at a kid playing and you figure they're just doing it because they're having fun, but they're understanding their world through these little adventures that they're having, these little experiments. It's often very nonlinear, but they're building a foundation of understanding of the world. And there's not really that much difference as far as adults go. A lot of scientists go in this linear direction, and there are times that that's the way to do things and that's very productive, but a lot of the most important discoveries that have been made have been off to the side on some tangent." ◼

Inquisitive kids may have more fun.

BACK TO BASICS

Can You Teach Science Using Rap?

From homage to Sir Mix-A-Lot ("I like Big Bang, and I cannot lie!") to a rap battle between Neil and a flat-Earth believer, rap readily lends itself to science. "It got me where I am now, says GZA, co-founder of Wu-Tang Clan, "because it all started with us studying lessons . . . Some of the lessons related to earth science, like 'What makes rain, hail, snow, earthquakes?'. . . The circumference of the planet, light traveling at 186,000 miles per second—these were [what] we studied coming up, so it kind of gave us an edge on being lyrical, with the wordplay, and flowing, just knowing things, word for word, and started us on our adventure to learn more about the universe."

Rapping isn't just about rhyming words. Building a knowledge base is a necessary first step, explains educator Dr. Christopher Emdin. "Analogy, metaphor, drawing connections, weaving stories . . . The more prolific the MCs, the more complex the lyrics are. The more complex the lyrics are is based on the ability to make connections that the layperson can't see, which is inherently scientific."

Are You Out of Your Mind?

Are All Drugs Dangerous?

TOUR GUIDE

What Can Marijuana Do for You?

I n the hands of an artisan, a sharp knife can turn a block of wood into a beautiful object. In the wrong hands, it can mean disaster. Similarly, drugs of all kinds can be good or bad—medicine or dope, says Cara Santa Maria, a science communicator. "Ecstasy, for example, which we think of as a recreational drug, has been used for couples therapy . . . There are lots of psychotropics that are prescribed. Anything from ADHD to schizophrenia to depression, anxiety—all of these disorders require drug treatment . . . with psychotropics."

An LSD tab featuring the Mad Hatter

"There are now studies . . . using ketamines for the treatment of depression," adds neuroscientist Dr. Heather Berlin. "[That] used to be a club drug called 'Special K.' If you take too much of it, you can disassociate; it's not good."

"Therapeutically, with well-trained and very elaborate supervision, it [Ecstasy] can be very helpful," says Dr. Mayim Bialik, neuroscientist and actress. ∎

With recreational use of marijuana now legal in parts of the United States, increasing attention is focusing on this psychotropic drug and its many effects. Users can experience very different results with its use. Just ask comedian Chuck Nice: "It can work both ways. It can make you go, 'Whoa,' when you shouldn't be doing that at all, or, it can take something that is already mind-blowing and just cause you to get deeper into that, like the fact that popping in and out of the multiverse on a quantum physics level would mean that we are the atom that's popping in and out, the whole universe is the atom that's popping—see, that is awesome!"

THINK ON THIS ▶ Where Did You Go on Your Trip?

"In the early 1960s, like a lot of people . . . I took a lot of drugs. I really wanted to see what I'd read about other forms of consciousness. To what extent would the world open for me? Would it reveal domains, perhaps, of natural or supernatural beauty or meaning? . . . I wrote my book *Migraine,* and I never took drugs again." —Dr. Oliver Sacks, neurologist and author

"My whole body below my neck went numb, and I just remember lying there, praying to God that I would be able to get up again and walk."

—CORY BOOKER, U.S. SENATOR

StarTalk Live! at the Apollo (Part 1)

Why Is Football So Dangerous?

I f it weren't for the violence, would American football still be a great game? The skill, strategy, and athleticism of the sport is beautiful, to be sure. But the devastating injuries to all body parts have become a dominant feature of the sport. The invisible wounds of football may be even more frightening; a well-meaning effort to protect the athletes decades ago appears to have led to the medical crisis of chronic traumatic encephalopathy (CTE).

"Football has the greatest concussion rate, and it's because we have the helmets," explains Dr. Ainissa Ramirez, materials scientist and co-author of *Newton's Football*. "The reason we had helmets to begin with is that people used to die from the game. They died because they had skull fractures. In the 1950s [face masks] became standard issue. What that did is change the way we tackle. We used to tackle with our shoulders; then we started tackling with our heads. By putting on the face mask, the helmet became weaponized. So that's what gave rise to concussions." ■

Football helmets have led to more lethal collisions.

BACK TO BASICS

What Can NASCAR Teach Us About Physics?

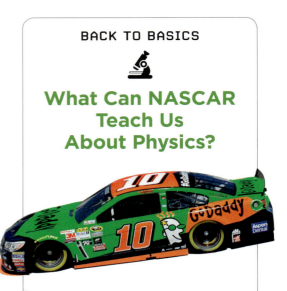

Every college physics major has been assigned at least one homework problem about auto racetracks. Physics manifests itself in just about every facet of auto racing, from the tilt angle of the track to turn speeds to fuel consumption to burning out the victory doughnuts. "If you are driving at the right speed for that bank, you do not have to turn the steering wheel; the track will turn you," says astro-steering-ist Dr. Neil deGrasse Tyson. "So as far as the car is concerned, it's going in a straight line. They don't ever have to turn the steering wheel to bank those turns . . . So that's a car driving in a straight line with the space-time continuum of the NASCAR track curving its path."

DID YOU KNOW

NASCAR vehicles routinely exceed 200 miles an hour in events at superspeedways like Daytona International Speedway, where the track bank angles range from 18 to 31 degrees.

How Do Roller Coasters Work?

Roller coasters do what their name implies: The wheels roll, and the cars coast—that is, the cars have no independent means of propulsion. All the motion comes from the conversion of the car's gravitational potential energy to kinetic energy and back again, over and over, as the car goes up and down.

The farther up from Earth's surface an object goes, the more gravitational potential energy it has. Once released, the object falls, and its potential energy starts to change into kinetic energy—the

"The highest point of a roller coaster essentially determines the fastest speed you will ever reach on that roller coaster, because it's all about energy."

—DR. NEIL DEGRASSE TYSON

steeper the drop, the faster the conversion goes. Throughout their ride, the cars must overcome the friction of the wheels, which will slow them down.

Finally, if the roller coaster has a vertical loop, the car has to be moving fast enough so that the centrifugal force it feels will pin the car to the track—even when it's upside down. The loops are also not circular—they're stretched vertically to have a steeper drop on the way down but also to make the car's ride smoother as it exits the loop. ■

A sculpted walkway in Duisburg, Germany, mimics a roller coaster.

Internet trolls hide in anonymity.

The Digital Revolution With Arianna Huffington

What Are Trolls, and Why Do They Exist?

The kind of troll you don't need to fear

nternet trolls use crude insults and foul language to turn online discussions into unpleasant bouts of usually anonymous electronic shouting. In many ways, trolls are like schoolyard and workplace bullies—and their behavior can range from poking fun at people to dangerous threats of violence. Publisher Arianna Huffington explains how to deal with them: "Trolls are incredibly ingenious, some of them. They want to basically have no other life except to circumvent this technology . . . There is something about anonymity that brings the worst out in people . . . Don't feed the trolls. Don't give them attention. Don't reward them. Don't laugh. Because you're giving them what they want." ■

DID YOU KNOW

In a survey of adult Internet users, 28 percent of the respondents admitted to researchers they had engaged in trolling—and most of them were men.

A Conversation With Nichelle Nichols

Can Television Fight Racism?

" I had a man at one of the *Star Trek* conventions in London, where they had the skinheads movement . . . Here was this man, at the front of the line, he couldn't have been more than 18 or 19, but he looked more frightening, with the skin head and the whole thing and the muscles and the tattoos, and he walked over to the table, and my security team tensed . . . and he said, 'I'm not really here as a fan, Miss Nichols. I just wanted you to know that because of you on *Star Trek,* I stopped being what I was on my way to be, and I came here dressed as I had been doing, so that you would know.' This guy is tearing up, and I'm looking up at him tearing, and he said, 'I can never be that again . . . I understand what the world and the future is, and *Star Trek* depicts it, and it's not what I was going through with all my pain and my life. I've done some dastardly things. I can only hope to make up for it.' And I stood, and I leaned across the table . . . and I said, 'Come here, son.' And I put my arms out, and tears came down his eyes, mine, and I held him. And I said, 'Don't look for forgiveness anywhere. God just forgave you. Your choice makes you whole.' And he thanked me, and he walked away, and I turned to my security, and they just stood there like their hands were just limp, and one man said, 'Well, I'll be damned.' And I said, 'No, you'll be blessed. You just got blessed.' "
—Nichelle Nichols, actress who played Lieutenant Uhura on *Star Trek* ∎

DID YOU KNOW
Hikaru Sulu was shown as the staff physicist in the pilot episode of *Star Trek* but served as helmsman throughout the rest of the series.

Star Trek's original cast in 1966

"That vision that [Gene Roddenberry, creator of Star Trek*] had: starship* Enterprise*, a metaphor for starship Earth. And the strength of the starship lay in its diversity, coming together and working in concert. Nobody was thinking that."*

—GEORGE TAKEI, ACTOR WHO PLAYED HIKARU SULU, HELMSMAN OF THE STARSHIP *ENTERPRISE* ON *STAR TREK*

"Don't you agree that astronomy and cosmology would look very different if you thought that the whole thing, the laws of physics, had been planned in advance . . . Isn't that a gigantic scientific fact, if it were true?"

—DR. RICHARD DAWKINS, EVOLUTIONARY BIOLOGIST

CHAPTER SIX

Is God Real?

Scholars of religion have suggested that belief in God—or the gods, or the ghosts of your ancestors, or any other powerful supernatural entity—helped human society evolve toward controlling the environment. If, for example, you wanted to prevent storms from sinking your boat, you prayed to God for calm seas. The all-powerful deity, pleased with your worship, would then spare you from danger.

Well, did your supplications really make a difference? Did God hear you? Heck, is God even there? It's not possible to test these questions scientifically. And today, with technology and good weather forecasting, is God even necessary anymore?

As long as we face great mysteries and great worries, the need for a scientifically unconfirmable God probably won't go away. More important, say social scientists, the purpose of God may be to give societies something in common to believe in—a guide to live by—an ideal to work toward. The need to be greater and do better than just oneself is, after all, a powerful part of being human.

Most religions include belief in a higher power.

Which Came First: the Big Bang or God?

In Western civilization, the fundamental philosophical idea that something had to have started everything else—that is, the universe—can be traced back to Plato and Aristotle. Centuries later, their ideas of the "uncaused cause" and the "prime mover" gradually mingled with religious ideas and joined contemporary concepts of "God" and "creation" in religions like Judaism, Christianity, and Islam.

The challenge of creation narratives of any kind is that, sooner or later, someone wonders what came before: What caused creation to happen? Did the universe—generally defined as "everything that exists"—have to exist before the first cause could exist? Or was the existence of the universe caused—in which case, something existed before everything existed?

"If there is a designer of the universe, that is a stupendous scientific fact—it's not something you can say, 'Well, we only think about that on Sunday.' "

—DR. RICHARD DAWKINS, EVOLUTIONARY BIOLOGIST

It's a circular argument that could go on forever. At least for now, we humans have to either accept the uncertainty of the cosmic origin or just choose to believe one of the many creation stories that we have conceived. Or both.

"I don't presume that everything has to have a cause," says Neil. "That's just how the world has manifested thus far . . . I'm open enough to other possibilities, that maybe the universe always was . . . The universe is under no obligation to make sense to us.

"Neither is God," points out Rev. James Martin. ∎

BIOGRAPHY
👓
Who—or What—Is God?

If God were not omnipotent and omniscient, God would have a serious identity crisis. The God of the biblical Old Testament is a single powerful being requiring worship from the Jews to the exclusion of all other possible deities, visiting divine wrath on any who strayed from that monotheistic path. In the New Testament, God takes three forms—the Father, the Son, and the Holy Ghost—but is still a single entity that grants mercy and absolution to anyone who seeks them. In the Book of Mormon, God, Jesus, and the Holy Ghost are three separate and distinct beings, united by a common purpose. Beyond these identities, there are thousands of other creators in myriad other traditions that are also said to have brought forth the universe.

THINK ON THIS ▶ Can You Test for God?

In one Old Testament story, Gideon is approached by God to lead the Israelites to battle. Gideon is skeptical and tests God's power by asking God to make a fleece wet overnight while the ground stayed dry, and then to keep the fleece dry overnight while the ground turned wet. This sorta-scientific test convinced Gideon, and he led the Israelites to victory. A more modern, more scientific test has yet to be constructed.

Cosmic Queries: Pseudoscience

What's the Story With the Creation Museum?

n May 2007 a group called Answers in Genesis opened the Creation Museum in Petersburg, Kentucky. After some 400,000 people visited the museum in its first year, attendance has steadily declined. "The problem is not what people believe. This is a free country. Believe what you want," Neil argues. "I will not tell you what to believe. What I will say to you is that if you want your belief, which is not based on objective truths (it's based on what are generally known as revealed truths: There is some sacred document that someone has truth revealed to them through, whatever forces that you recognize in your religion) . . . if those truths conflict with objectively verifiable truths, and you want to teach that as science, that's the beginning of the end of the technological foundation of your culture." ■

An imagined prehistoric scene at the Creation Museum in Kentucky

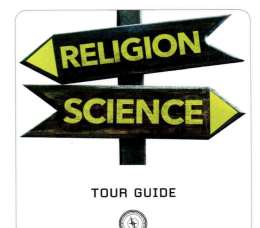

TOUR GUIDE

Can You Have a Rational Argument With a Creationist?

Scientific research shows that people who hold irrational beliefs are very difficult to dissuade with rational arguments. "If people think that the first woman was made by God putting Adam in a trance, then taking a rib from a side—if you believe that, there's no argument," says naturalist Sir David Attenborough. "I mean, there is no rational argument."

Sometimes, irrational beliefs are beneficial. "True love conquers all" may not be rational, but it sure is a nice thing when it happens. Other times, though, there are clearly negative consequences, especially in science. Neil explains: "The problem is when people who want to learn science think that creationism is science. They have been removed from the frontier of cosmic discovery." Irrational beliefs are, not surprisingly, usually best countered by irrational arguments—such as emotional appeals.

The Value of Science With Brian Cox

What's Magic About Magical Thinking?

Horoscopes. Superstitions. Santa Claus. The tooth fairy. It's fun to indulge in magical thinking—but often we don't recognize it as such. As much as we like to believe we are ruled by reason, most of our behavior is rooted in much more basic impulses, including biological needs and emotions—and thus not at all reasonable. So what can we do to fend off magical thinking?

Neil has one answer: "Science literacy, just as a state of mind, is quite the vaccine against those who speak in pseudoscientific ways . . . The human brain is awful at interpreting what it experiences . . . Our susceptibility as humans to cognitive failure is extraordinary . . . We invent methods and tools to reduce our susceptibility because we're honest with ourselves about it. That is what science is. That's what the scientific method is, at its heart: Do whatever it takes to not fool yourself into thinking that one thing is true when another thing is true."

"Science books start 'Well, of course, we may be wrong,'" adds physicist Dr. Brian Cox. "That's implicit in every science book . . . Imagine if every book, if every philosophy, if every religious document began with that." ■

DID YOU KNOW

We now know the age of both Earth (4.54 billion years) and the universe (13.8 billion years) to an accuracy of better than one percent—more precisely than most people know the ages of their friends and neighbors.

THINK ON THIS ▶ How Was David Able to Kill Goliath?

In the Bible, the shepherd boy David killed the Philistine giant Goliath using a sling. As it turns out, slings can be potent missile weapons—a shot to the head could be lethal. "David has superior technology, and he's up against a blind man," says author Malcolm Gladwell. "So let's just be clear about who the underdog is in this particular situation."

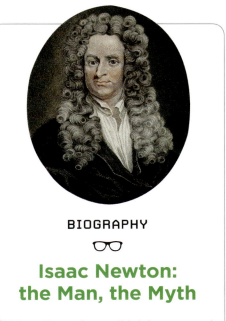

@neiltyson: "On this day long ago, a child was born who, by age 30, would transform the world. Happy Birthday Isaac Newton b. Dec 25, 1642."

Cosmic Queries: Mayan Apocalypse and Other Disasters

Why Did Pope Gregory Change the Calendar?

The Julian calendar, named after Julius Caesar, was based on ancient Roman astronomical calculations. It worked well for centuries, but it was just slightly out of sync with the orbit of Earth around the sun, and the error accumulated as the years went by. In 1582, Pope Gregory XIII introduced a new calendar, which, with only very minor adjustments, is the one we use today.

"Pope Gregory was very concerned about this, because the first day of spring was migrating backwards in the calendar. It turned out Easter, which was defined as the first Sunday after the equinox, was at risk of landing on Passover, where you would have overlapping rituals between Christians and Jews," Dr. Neil deGrasse Tyson, astro-calendar-ist, points out. "And the Pope said, 'We'll have none of that.' So they had to change the calendar in such a way so that they would never overlap again. That was his motivation. It wasn't just, 'Oh, I have this cosmic need to keep correct time'; it was 'I don't want Easter to look like Passover.'" ■

On the Julian calendar, the New Year in England began on March 25.

BIOGRAPHY

Isaac Newton: the Man, the Myth

Sir Isaac Newton's actual birthday, as carved on his tombstone in Westminster Abbey, was December 25, 1642. In his time, England still used the Julian calendar; so if we go back and change his birthday to the Gregorian calendar reckoning, his birthday would have been January 4, 1643. Historians, though, don't usually make that kind of adjustment for people born before the changeover. Not that it matters, really. No matter what the date of his birthday, Isaac was a brilliant man, who discovered an entire branch of mathematics, invented a new kind of telescope, developed modern physics, and even helped design British coins, as warden of the royal mint. He was, by all accounts, difficult to get along with, but his friends and supporters helped make sure his tremendous intellectual gifts were shared with—and transformed—the world.

"The thing is, Santa had something else that I don't think Isaac Newton had: He's magic! And that is a huge time-saver when you're gonna go worldwide."

—BILL NYE THE SANTA GUY

Why Did Einstein Use the Word "God" So Much?

Both religious and antireligious people have wondered about Albert Einstein's frequent references to God in his nonscientific writings. Some think that might have done science a disservice. "Einstein unfortunately muddied the issue by using the word 'God' rather freely, and people therefore want to claim Einstein [as religious]," evolutionary biologist Dr. Richard Dawkins says. "Einstein used God as a metaphor, and he said things like, 'What I really want to know is: Did God have a choice in creating the universe?' He simply meant, 'Is there only one way for a universe to be?'"

As it turns out, Albert Einstein valued his Jewish heritage and considered himself agnostic, with neither favor nor animosity toward religion. He rejected the idea that science and religion are incompatible. Indeed, he wrote for the *New York Times Magazine* in 1930, "Science without religion is lame; religion without science is blind." ∎

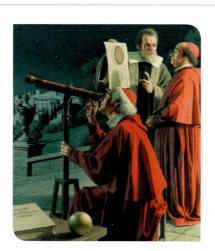

BIOGRAPHY

Who Was Galileo Galilei?

Considered by many the first scientist, Galileo Galilei (1564–1642) was also a great writer and science communicator. He argued that truths about the natural world must be based on observation, not just philosophy. He performed the first experiments on gravity, supposedly dropping cannonballs off the Leaning Tower of Pisa. He was brought before the Inquisition on charges of heresy for showing that Earth moves around the Sun. Even after that, under house arrest, he wrote the book that started the theoretical study of physics—work completed by Isaac Newton (who, by the way, was born the same year Galileo died).

Einstein extolled the universe's mysteries through physics.

"If it were as simple as saying all the smart people are atheists and all the stupid people are 'religulous,' you know, it would be very simple. But it's not that simple."

—BILL MAHER, COMEDIAN AND TV HOST, WRITER AND STAR OF *RELIGULOUS*

Does Religion Rise From Rational or Irrational Thought?

RICHARD DAWKINS: It may be that you do need a certain amount of illogic . . . If you see a . . . rustling in the trees, it could be a leopard . . . but it's much more likely to be the wind . . . But when your survival depends upon . . . the rather lower probability that it might be a leopard, the prudent thing is to be more risk-averse than a scientist probably would be. ■

REV. JAMES R. MARTIN: I think religion is based on people's experience of God. And religion is a way of relating to God and relating to one another, and through that, to God . . . Reason and faith are not inconsistent; logic and faith are not inconsistent. But there are some times that you might think illogically . . . like falling in love. ■

"I dozed off in the lounge, woke up about 40 minutes later to a commotion, went into the next room and saw the image of the tower after the first plane hit."

—SETH MACFARLANE, COMEDY WRITER

TOUR GUIDE

Miracle or Coincidence?

The comedy writer and television producer Seth MacFarlane was scheduled to take one of the flights that crashed on September 11, 2001. Did his near miss lead to a religious experience? Here's how he frames what happened: "I realized, "Oh my God, that was the plane I was supposed to be on . . . . I was delighted to find, not on that day, but much later . . . that my rational convictions were pretty much intact after that happened. When I really sat down and thought about it, I thought, Well, you know, I've missed flights before. I've missed a lot of flights before, for being late . . . And then also, when you think about it, there's people who miss every flight that takes off . . . There's no need for a radical alteration of my life's philosophy."

THINK ON THIS ▶ How Do Otherwise Rational People . . . ?

"How can otherwise intelligent people believe in a talking snake? How do people build this wall in their mind between what they must know, in part of their mind, is untrue—and yet they maintain this belief? That, to me, is the most fascinating part of it . . . . And no one will ever really answer it. But it's fun to try to find out, and it made for great comedy." —Bill Maher, comedian and TV host, writer and star of *Religulous*

The Digital Revolution With Arianna Huffington

Does Science Negate Religion?

History is full of examples of scientists who are religious and theologians who are scientific. The Vatican Observatory, for example, is a research institution run by deeply religious Jesuit clergymen who teach graduate courses, run an advanced telescope in Arizona, and host astrophysics summer schools each year.

"I asked our editor to focus a lot on this intersection between science and religion, and to break the illusion that scientists are all antispiritual, and that science negates spirituality, which, of course, it doesn't," says publisher Arianna Huffington. "I find the best scientists are very humble, because even though they discover so much, they are always aware of how much is left to be discovered. And for me, that's probably the essence of religion: that we don't really know a lot of what life . . . and the universe is about."

Most religious scientists, interestingly, do not believe in literal interpretations of ancient documents. Indeed, in the early 17th century, Galileo Galilei quoted his friend, the Roman Catholic cardinal Cesare Baronio: "The Bible teaches us how to go to heaven, not how the heavens go." ∎

TOUR GUIDE

Can We Talk to Trees?

HIS HOLINESS THE GYALWANG DRUKPA: Trees will talk to you. Plants will talk to you. Nature needs to be respected . . . Not only from the point of view of Buddhism or religion, Christianity or something . . . We just talk about the reality. So this is the religion; this is what we should be: We should always be with the reality. This is what I'm thinking. I don't know. You should correct me again.

EUGENE MIRMAN: Yes, Neil, you should correct him. First of all, trees are good listeners, but not good talkers.

JASON SUDEIKIS: Set him straight, Neil. Put him in his place.

NEIL: Your Holiness, there is nothing I can add or subtract from what you just said.

THINK ON THIS ▶ Does Belief Reflect Upon Intellect?

Bill Maher has a strong opinion about religious belief: "If you believe in Santa Claus, a god, Jesus, whatever you want . . . I really have to disqualify you from the highest rank of thinkers . . . And I don't even put myself in the highest rank of thinkers. I'm not saying, 'Oh, I'm up here in the pantheon, and you're not.' I'm just saying, 'I can't quite go there with you.'"

Do You Believe in Angels?

suspect that visionary experience, whether drug-induced or in dreams or whatever, has played a part in the genesis of every-thing from folklore to religion," said neurologist and author Oliver Sacks. "For example, for some reason—there are physiological reasons for this—Lilliputian hallucinations, so-called, are rather common: [visualizations] of little people. And one finds in almost every culture that there are elves, fairies, trolls, little people. One wants to say they're not at the sort of lofty level of angels in the heavens, but they do represent another reality. I think this is almost built into the brain as well as built into culture."

Dr. Sacks was definitely onto something. We all hallucinate at night in our dreams, and sometimes we remember them after we wake up. Medically speaking, it's perfectly normal for some of those dream-making mechanisms to produce unreal sensory experiences while we're awake. Our conscious minds could indeed contextualize them as conversations with angels—or whatever else makes "sense." ∎

BACK TO BASICS

A Mission From God?

In the classic 1980 movie *The Blues Brothers*, the brothers' car—the Bluesmobile—is implied to have magical powers because they are on a godly mission. (A scene where the car gets "charged" with power in an electrical substation was deleted for length.) Dan Aykroyd, the movie's coscreenwriter, who also played Elwood Blues, put it this way: "Once we undertook the mission from God, we had the power of the Godhead behind us. We had divine power behind us. We had the power of belief. We had the power of a universal sense of the world and a shared purpose. So I think the car took on divine attributes. It became a holy relic at that point."

Fresco in the Shepherds' Field church, Bethlehem

DID YOU KNOW
In *The Blues Brothers* (1980), 60 police cars were wrecked for the film—a record at the time for most cars smashed for a movie.

Who You Gonna Call to Bust Your Ghosts?

n 1998 British engineer and part-time psychic investigator Vic Tandy published a study showing that exposure to infrasound—sound waves below the threshold of human hearing—could cause eerie discomfort and possibly ghost-like visual hallucinations in people. Might further investigation of this sort of phenomenon be worthwhile?

Dan Aykroyd, actor and coscreenwriter of *Ghostbusters* thinks it is: "I want scientists to begin to inquire: What is the particulate matter? What is going on electrically? What's acting here? Is it oxygen? Nitrogen? What's in the air that can produce an apparition where people can actually manifest, in their vision, the apparition of someone who's dead? That's where I think spiritualism should go, right into the scientific. And so we've got to entice serious scientific inquirers to look into this. So far, no one is interested—except me." ∎

> "If you're alive and then you're dead, and you're on a scale, your weight doesn't change . . . As for a soul—if you want to believe there is one—there is no evidence that it has mass."
>
> —DR. NEIL DEGRASSE TYSON, ASTRO-SOUL-ICIST

The battered cast of Ghostbusters *in 1984*

DID YOU KNOW
Around the year 1900, numerous scientists tried to take pictures of the human soul using x-rays, a newfangled technology at the time. It didn't work.

THINK ON THIS ▶ Is Hell Located Inside a Black Hole?

A few years ago, two televangelists on a U.S. TV broadcast claimed to show that black holes fulfill all the technical requirements of hell as described in the Bible. They were awarded an Ig Nobel Prize for their efforts. Inserting scientific ideas into religious contexts is hardly unique; another calculation was made, for example, showing that heaven must be hotter than hell. And in 1951, Pope Pius XII issued a papal encyclical declaring that the big bang is proof that God exists.

How Do We Know We Landed on the Moon?

Hundreds of hours of video, thousands of photographs, tens of thousands of people who worked on the Apollo project, and millions of pages of documents are not sufficient evidence to convince some people that humans have walked on the Moon. "One really neat piece of evidence is to look at the video of Buzz running on the Moon and the video from the later flights and the movies, and watch the way the dust moves," author Andrew Chaikin tells us. "It's not like anything you've seen on Earth . . . and that's impossible to fake—and certainly was impossible to fake in the 1960s and '70s."

And how about talking to one of those humans who took a lunar stroll?

"You put your foot down, kick your foot, and here on Earth it just kind of moves stuff out in front," astronaut Dr. Buzz Aldrin says. "But on the Moon, you do that, and it goes out and it all forms in a semicircle. It really is different, because there's no air." ■

BACK TO BASICS

Why Are People Fascinated With Crystals?

With their unusual shapes, crisp edges, and flat planes, large mineral crystals form only under special geological conditions. Not surprisingly, some people imagine they may have supernatural properties—but even if they don't, they can still be beautiful. Astro-crystal-ist Dr. Neil deGrasse Tyson explains how crystals got their power: "Crystals were the world's first transparent solid things. Think about that. Normally, you think of solid, and nothing goes through it. Now I've got a solid thing that light can pass through. That's an awesome thought, and there it is in nature. Crystals were valued simply because they were different, and they were transparent, and it's a natural part of human curiosity to pick up that which is different and bring it home."

A man adjusts a display at Kennedy Space Center, Florida.

"Those are the seeds of the birth of science: the fact that we are curious about some things that are different than others."

—DR. NEIL DEGRASSE TYSON

FUTURES IMAGINED

Zombies! Superheroes! Warp drive! Aliens! And, maybe most fantastical of all, time travel! As Albert Einstein wrote: "Imagination embraces the entire world, stimulating progress, giving birth to evolution. It is, strictly speaking, a real factor in scientific research." We can't always create what we imagine. But we also can't intentionally create what we don't imagine first. Science fiction looks at the best and the worst of humanity—the endless possibilities of Earth and space—and that mind's eye will forever push the boundaries of "what is" toward "what could be." So let's go to the limit—and then zip on by!

"It's their biological imperative to spread the virus. Through the eating, they're not ingesting nutrition, but it is an act that is familiar to their DNA. They already know how to eat, and that's the best way to spread the virus. They are a walking plague. They are literally a virus."

—MAX BROOKS, AUTHOR OF *WORLD WAR Z*

CHAPTER ONE

When Are the Zombies Coming?

Mindless. Unstoppable. Terrifying! Zombies are all the rage, taking over not only in horror films (*28 Days Later*) and on TV (*The Walking Dead*) but also sci-fi conventions (*Night of the Living Trekkies*), math class (*Zombies and Calculus*), and Victorian literature (*Pride and Prejudice and Zombies*).

Modern zombies are fictional, but the zombie phenomenon is really an allegory about plagues, death, and the end of civilized society. And the microscopic pathogens that can cause such disasters are all too real. As technology advances into medicine, it seems all too likely that we could sow the seeds of our own destruction, either biologically with experiments on viruses and germs, or mechanically with artificial nano-invaders.

Don't despair. Through scientific research and medicine, we are starting to peer into the fiction and extract the reality. And with that knowledge will come power—the ability to eradicate disease, and to harness viruses to do much more good than harm.

Zombies have been imagined many ways but are nearly always murderous and bloodthirsty.

Zombie Apocalypse (Part 2)

Could Zombies Actually Exist?

Humanity's view of zombies has changed over the years. In fantastical antiquity, as depicted in swords-and-sorcery games like Dungeons & Dragons, a zombie was a magical "undead" creature, similar to a vampire or lich. About a century ago, popular accounts of zombies in Haiti described them as reanimated human corpses controlled by evil voodoo necromancers—or, alternatively, as "zombified" living people under the influence of psychoactive drugs. Science-fiction zombies entered the mainstream with Hollywood B movies like *Plan 9 From Outer Space* (1959, considered by some to be the worst film ever made) and *Night of the Living Dead* (1968).

> *"When something bites you, you don't turn into that thing . . . If that were remotely possible, Evander Holyfield would have turned into Mike Tyson years ago."*
>
> —DR. NEIL DEGRASSE TYSON, ASTRO-NEVER-HEARD-OF-WEREWOLVES-IST

Today's fictional zombies are biological, created by an infectious virus, and are even deadlier than those of decades ago. They're not only hard to kill, but they have an insatiable appetite for human flesh, and their bite would infect you with their zombieness—ensuring that you'll wind up just like them upon your demise. From video games like *Resident Evil* to novels like *World War Z,* each zombie depiction is more vicious and bloodthirsty than the last. We know that some dreaded diseases, like rabies, can be passed along through saliva, but it takes weeks for symptoms to develop, not seconds. So while some kind of zombie germ may be conceivable, it's a stretch. ∎

TOUR GUIDE

What Are the Best Weapons for a Zombie Apocalypse?

Max Brooks, author of zombie apocalypse novel *World War Z*, says that fighting zombies is about finding the most energy-efficient way to kill them: Do not engage, but evade and survive. His weapons of choice:

GUN

"Every time you pull a trigger, where is the next bullet going to come from? Who's making bullets?"

BLADE

"Machetes are the best. A shovel works just fine—a good hand weapon."

BOW AND ARROW

"You can't just pick up a bow and arrow and go Robin Hood . . . You better start practicing right now."

THINK ON THIS ▶ What Do Zombie Fantasies Say About Society?

"I think the reason the zombie craze is so crazy right now is because we're living in such anxiety-ridden times . . . It's the one apocalyptic scenario in which we can feel empowered. You cannot hit a credit default swap in the head . . . If you keep a cool head and, as the British say, keep calm and carry on, you'll be all right."
—Max Brooks, author of *World War Z*

Cosmic Queries: Viruses, Outbreaks, and Pandemics

Is There Really a Zombie Fungus?

t's true. There's a zombie fungus on the loose! Sort of. "I don't know that we've actually seen a fungus that turned people into zombies, but we certainly see fungi that execute all kinds of clever activities to force the behavior of whatever they infect to facilitate spread of their spores," says science journalist Laurie Garrett. "You might argue that's what athlete's foot in a gym is."

There are indeed parasitic fungi out there: Alfred Russel Wallace, the famous British naturalist who pioneered the theory of evolution by natural selection with Charles Darwin, discovered *Ophiocordyceps unilateralis,* a parasitic fungus that infects tropical tree ants. The infected ant convulses, falls to the jungle floor, and climbs up back toward the treetops. On its way back up, it bites down hard on a leaf, and its mandibles lock in place. Over several days' time, the ant dies. The fungus sprouts through the dead ant's head, and new spores are released to seek their next victims. ∎

BACK TO BASICS

How to Cure Zombies With Salt

Author Mark Kurlansky talked about the treatment of zombies in Haiti: "I have spent a lot of time working in Haiti, so I already knew that salt was used to cure zombies . . . because salt takes away evil . . . If somebody has been zombified, you can bring them back to normal with salt . . . I have to say: I haven't done it. But there is this association with salt preventing evil and curing evil, because it stops rotting."

The Cordyceps *fungus sprouts from a murdered ant.*

"You've heard the phrase 'cat lady'—somebody who owns hundreds of cats and behaves in a very eccentric way? Maybe a lot of those people are infected with this parasite."

—MARC ABRAHAMS, FOUNDER OF THE IG NOBEL PRIZES, ON THE PARASITE *TOXOPLASMA GONDII*

Zombie Apocalypse (Part 1)

How Would a Zombie Virus Spread?

Zombies are particularly wrathful in large numbers.

A virus is merely a bundle of genetic material with an imperative to make copies of itself. Epidemiologists look to both biological disease vectors (like air, water, or blood) and human behavior to make models of how diseases spread.

Max Brooks, author of *World War Z*, says he based his zombie virus on AIDS: "I wanted to make it very hard to get, just like AIDS was very hard to get. And therefore, from a storytelling point of view, the mistakes were made by us. Because the truth is, let's face it, if in 1980 Reagan had gone on TV and said, 'My fellow Americans, there's a disease that's real hard to get, but if you get it, it's going to be really bad. Here's 10 things you can do to avoid it'—boom!—AIDS would have been a paragraph in a medical journal . . . We could have made AIDS extinct with a pamphlet; that's how we could have stopped it."

"If only that were true," argues epidemiologist Dr. Ian Lipkin. "I don't think a pamphlet would have changed the course." ∎

> "What I love is that you're starting to get genuine thinkers; you're starting to get genuine academics and smart people who are really looking at a zombie plague from an academic point of view."
>
> —MAX BROOKS, AUTHOR OF *WORLD WAR Z*

DID YOU KNOW

In his short story "The Giving Plague," astrophysicist Dr. David Brin describes a virus that compels people to donate blood—it's the only way the virus can spread from one host to another.

LAUGH OUT LOUD ▶ **With the U.S. Centers for Disease Control**

"Get a Kit—Make a Plan—Be Prepared" reads the poster, with a zombie staring menacingly outward. In 2011, Dr. Ali Khan at the Centers for Disease Control led the distribution of tongue-in-cheek online antizombie materials, including the graphic novel *Preparedness 101: Zombie Pandemic,* to engage teenagers and children with the topic of emergency planning, without getting too serious about disaster and disease. It spurred the largest amount of Internet traffic the CDC has ever had.

PREPAREDNESS 101:
ZOMBIE PANDEMIC

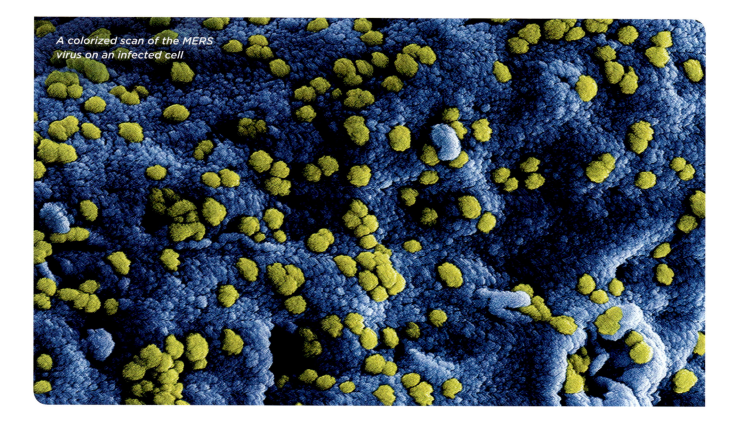

A colorized scan of the MERS virus on an infected cell

How Do Viruses Really Work?

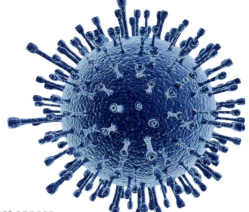

Viruses that appear spherical are actually icosahedrons.

Though epidemiologist Dr. Ian Lipkin says "a very famous virologist once referred to a virus as a piece of bad news wrapped up in a protein coat," it's unlikely viruses have any conception at all of good or evil. In fact, not all viruses are bad. But they do all work the same: A virus will inject itself into a cell and then hijack its machinery, forcing the cell to make the genetic material and proteins needed to replicate the virus. In other words, a virus can turn a cell into its slave—or zombie!

If the cell's internal immune response isn't strong enough to resist the virus, the cell will die once the virus is done with it. From there, the viral copies will disperse to find the next cells to enslave. And so the virus continues to live, and spread, one cell at a time. ■

"It's evil beef jerky."
—COMEDIAN EUGENE MIRMAN ON VIRUSES

Zombie Viruses or Real Viruses: What's the Bigger Threat?

In fiction, zombie pandemics are almost always caused by poor behavior or planning on the part of healthy humans. So how does a zombie-virus attack stack up to the thousands—maybe millions—of kinds of real viruses on Earth?

IIIIIII
▼ INCUBATION

Pathogens incubate, or reproduce, for varying periods of time without the host's showing symptoms. The longer the incubation period lasts, the more people can be infected before an epidemic is noticed. The zombie virus usually incubates quite quickly.

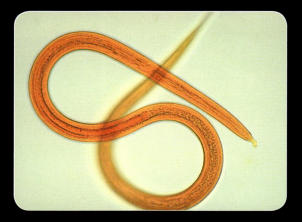

IIIIIII
▲ CONTAGION

Viruses like Ebola can travel several feet via droplets in the air—when, for example, an infected patient coughs up blood. The zombie virus can be transferred only by direct contact and the exchange of bodily fluids.

IIIIIII
◀ TRANSMISSION

Viruses like measles or influenza can travel through the air for long distances and linger in the air for long periods of time. As long as a zombie doesn't touch you, you're fine.

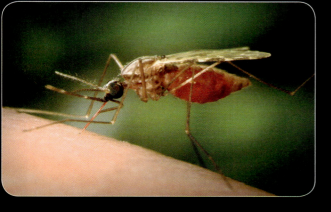

◀ VECTORS

Viruses like dengue fever and Zika are transmitted by mosquitoes—which can travel a long way and carry the virus with them. Zombie viruses can usually only be passed along through humans.

|||||||

▶ MORTALITY

If everyone infected by a virus dies from it, an epidemic could be self-limiting, because there'd be no more carriers left. The zombie virus lives on, though, even after its host dies, so it won't self-limit.

|||||||

▲ FATALITIES

In approximately one year, a MERS epidemic killed about 600 people; SARS, 800 people; Ebola, 11,300 people; the 1918 influenza, about 50 million people. The death toll from all zombie viruses so far: zero.

AND THE WINNER IS . . .

The Top Three Viruses

"If you ask me what bothers me most in terms of the state of the world, it really is HIV," says Dr. Ian Lipkin, epidemiologist. "Number two would be influenza. And number three would be the one I don't know about, because we're seeing new things all the time."

HIV, the virus that causes AIDS, can be transmitted only through direct contact with bodily fluids. Yet because of human behavior, it has reduced life expectancy in some African nations by more than a decade, and it continues to spread rampantly worldwide through unprotected sexual activity and the sharing of infected needles. As for influenza, it mutates so quickly that it continues to kill hundreds of thousands of people each year. ■

Zombie Apocalypse (Part 2)

What Does Climate Change Have to Do With Viral Outbreaks?

Global warming changes the boundaries of climate zones worldwide, allowing organisms to thrive where they have not survived before. Meanwhile, humans are changing the boundaries of habitats worldwide as jungle or forest or swamp is transformed for living and farming. These shifting borders create more interactions between humans and other species, and more opportunities for viruses to jump from those species to us. Researchers today think, for example, that AIDS originated in west equatorial Africa many years ago, when humans came into contact with chimpanzees that harbored the virus.

"Wherever climate change and habitat intrusion is occurring as a nexus, we see species put under tremendous pressure that carry viruses that humans have not been exposed to," explains science journalist Laurie Garrett. "Almost all of the big epidemics that we've seen in the last decade-plus have come from fruit bats that normally pollinate the rain forest. And as the rain forest is under stress and the upper canopies are getting overheated, desperate bat populations are moving closer and closer into human areas and passing their viruses to our livestock, and eventually to us. And guess what that includes? SARS. Ebola turns out to be a bat virus. Marburg turns out to be. Lyssa. Hendra . . . And we're unable to really predict or quantify the risk." ∎

DRINK OF THE EVENING

The Parasitic Poison

Concocted by Dr. Neil deGrasse Tyson and Kim, the bartender at the Bell House.

1 oz. Crystal Head Vodka (note: the brand is critical)
A kiss of silver tequila
3 dashes of bitters
1 oz. ginger beer
A little lemon
A cherry

Pour into tumbler over ice and bruise.

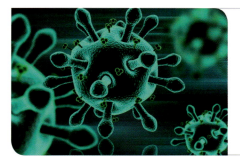

THINK ON THIS ▸ How Do We Stop a Global Viral Outbreak?

Plagues often start among groups of poverty-stricken and marginalized people, says Laurie Garrett, a science journalist. "A whole lot of epidemics are simply because of reused syringes in poor country hospitals."

The best solution for the HIV/AIDS epidemic so far, in places from San Francisco to South Africa, appears to be to work with such populations, sometimes one person at a time, to slow or stop the spread of the virus.

Zombie Apocalypse (Part 2)

How Did SARS Spread Around the World?

"Fear does kill in an epidemic. People behave really stupidly and put themselves at great risk," says science journalist Laurie Garrett. "[During] the SARS epidemic, I was in Hong Kong, and there were two major hospitals that the bulk of all of the patients went to . . . It started in China and was covered up. It started in November 2002, [and] we didn't know about it until . . . it hit Hong Kong. And it hit Hong Kong because one individual was infected, was terrified, knew what was going down, because he was a doctor, staggered across the border, went to a hotel called the Metropole in downtown Hong Kong, stayed on the ninth floor, and everybody else on the ninth floor that pressed the 9 button on the elevator got his disease. They were travelers. They went to their respective airport destinations and took the virus—to Vietnam, to Toronto . . ." ■

A factory worker in Beijing dons a mask to ward off SARS in 2003.

BACK TO BASICS

Can DNA Jump Between Species?

A research project by plant geneticists recently showed that the genome of the common sweet potato contains DNA from a soil bacterium that causes abnormal growths at the root crowns of many trees and shrubs. This isn't the kind of DNA transfer from typical hybridization, but rather much more like naturally occurring genetic modification—GMOs without human intervention.

How might the bacterial DNA have entered into the sweet potato genome? Possibly a virus attacked a sweet potato cell long ago, injecting all kinds of foreign DNA—but the cell survived, thanks to its immune system, and the DNA was incorporated into the cell.

DID YOU KNOW
Modern gene therapy, being developed by doctors to fight cancer and other serious diseases, involves the genetic modification of human cells by introducing outside DNA, sometimes from viruses, and also using viruses to deliver the genetic material.

"I went to the president [of Ghana] three times, and I told him we were going to change the name of the guinea worm to 'Ghana worm.'"

—JIMMY CARTER, 39TH PRESIDENT OF THE UNITED STATES

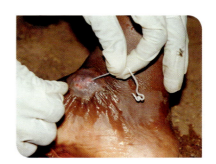

Combating Disease With Jimmy Carter

Is Politics the Best Weapon to Fight Outbreaks?

Jimmy Carter, the 39th president of the United States, called on his diplomatic skills to help eradicate guinea worm disease. Caused by a tropical parasite, the disease infects a person who has been drinking water containing the worm's larvae, and results in debilitating symptoms. President Carter's efforts through the Carter Center carried him to numerous countries, including Sudan during a civil war.

Here's what President Carter has to say about his time in Sudan: "I went there and negotiated for quite a while with south and north, and finally they agreed on a cease-fire, just so we could

do away with guinea worms both in the north and the south . . . They still call it 'the Guinea Worm Cease-Fire.' . . . They quit fighting for more than six months . . . It shows that if you give people a chance in a very poverty-stricken country to correct their own problems, they do it superbly." ∎

U.S. president Jimmy Carter

TOUR GUIDE

Is It Genocide to Wipe Out the Guinea Worm?

Only 22 cases of dracunculiasis, or guinea worm disease, were reported worldwide in 2015. Since human hosts are a key part of the guinea worm's life cycle, it may soon be possible for the guinea worm to be completely eradicated from the world. Should we hesitate to destroy an entire species?

"I'm very sentimental about parasites, but I'm sentimental about my eight-year-old daughter, too," says biologist Dr. Mark Siddall. "I think that if there's a parasite that goes extinct if we go extinct, then we don't have a moral responsibility to save it . . . For the guinea worm to not go extinct, it'd require that we assign someone's child to carry it. Is that going to be your child? Is it going to be mine? And are we going to do that out of some weird sense of ecological guilt?"

DID YOU KNOW

Guinea worm larvae infect water fleas in central Africa. When humans drink infested water, the guinea worm grows inside the human body for about a year, to a length of two to three feet, and then eats its way out of a leg or foot over a period of several weeks.

Combating Disease With Jimmy Carter

Didn't We Already Wipe Out Polio?

P olio was one the most feared diseases of the 20th century. This highly contagious viral disease struck without warning, rich and poor alike, and it paralyzed, crippled, and killed hundreds of thousands of people each year. Then, Drs. Jonas Salk and Albert Sabin developed polio vaccines; within a generation, the number of cases in the United States dropped to almost zero.

Today, thanks to an ongoing eradication program, the number of reported cases worldwide dropped from nearly 400,000 in 1988 to barely 100 in 2015. We're not done yet, though. Invertebrate biologist Dr. Mark Siddall explains why: "Right now there's fewer than 500 new cases of polio, but they're in Waziristan, they're in Syria, they're in southern Somalia, and they're in northern Nigeria. These are places of conflict . . . What goes on in a war-torn area is an inability to provide services, and an inability to track cases and find out where they are. Both of those, at the front end and the back end: You lose the connection to health care, and it's devastating." ∎

> "The biggest lesson from tuberculosis, HIV, go down the list: You don't want to be part of a disenfranchised group or a group that the larger society looks down upon. Because if you are that individual, society will not be there for you."
> —LAURIE GARRETT, SCIENCE JOURNALIST

> "By understanding the science of disease . . . we've been able to preserve the lives of millions of people around the world, raising their quality of life and making them more productive, so that people everywhere enjoy longer, healthier lives. It's wonderful."
> —BILL NYE THE PUBLIC HEALTH GUY

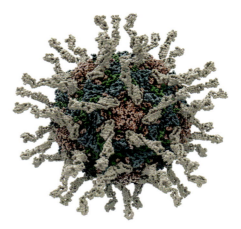

Protruding proteins of the polio virus seek out a host.

DID YOU KNOW

If the global program of polio vaccination and eradication stopped today, the annual number of polio cases worldwide could skyrocket to 200,000 in a single decade. The program costs more than one billion dollars a year to operate.

THINK ON THIS ▶ Did the CIA Use a Vaccine to Hunt Bin Laden?

The U.S. Central Intelligence Agency used a fake vaccination campaign to search for Osama bin Laden in Pakistan. "It was a fake hepatitis vaccine campaign, [and they] never did actually access the bin Laden kids," says science journalist Laurie Garrett. "Nevertheless, the story has gone wild, and now Islamists—Taliban, al Qaeda, al Qaeda offshoots—are slaughtering polio workers in Pakistan, in Afghanistan, in Yemen . . . in Nigeria . . . It's open season."

Zombie Apocalypse (Part 2)

How Dangerous Is the Anti-Vaccine Movement?

From November 2012 to July 2013, more than 1,400 cases of measles were reported in Wales. Of those, 1,200 were in the area of Swansea; one man died. It turned out that, 10 years earlier, vaccination coverage in Wales had fallen below 80 percent, and below 70 percent in Swansea—low enough that the disease returned with a vengeance. Science journalist Laurie Garrett explains what happened: "They've come to the conclusion, which has absolutely no basis in scientific reality, that their babies will have autism because of a vaccine . . . It started with a guy in the U.K. named Andrew Wakefield, who claimed to be able to prove this, and it turned out to be totally bogus . . . yet it continues to resonate in a certain kind of conspiratorial way." ■

Protesters at the U.S. Capitol demand a ban on mercury in vaccines.

BACK TO BASICS

What Should We Really Be Afraid Of?

"If you're a big living thing, like a human, you might think your enemies are lions and tigers and bears, oh my . . . No, it's tiny things, like germs and parasites. They've wiped out whole societies and civilizations, with things like the bubonic plague, Ebola, and the flu. And who knows what else is out there . . . just waiting to come and get us. You need a microscope and special skills even to see them . . . That's why so many people around the world have trouble accepting how dangerous they can be."
—Bill Nye the Infection Guy

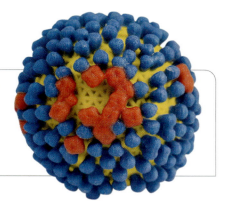

THINK ON THIS ▶ Could Researchers Create a Viral Apocalypse?

Could we learn about an epidemic by creating one?

LAURIE GARRETT: Groups are now saying, "Well, why can't we [make] the very organisms that would create the worst-case epidemics and then study them in the labs?"

EUGENE MIRMAN: That sounds exactly like the end-of-the-world scenario beginning of every movie.

Cosmic Queries: Viruses, Outbreaks, and Pandemics

Could Self-Aware Nanobots Turn People Into Viruses?

"The question to ask is, 'Is there a way to make a nanobot self-reproducing?' If nanobots could be self-reproducing, then indeed you could have an out-of-control infectious problem."

—LAURIE GARRETT, SCIENCE JOURNALIST

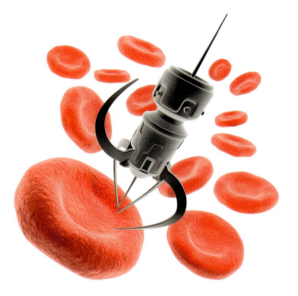

An imagined nanobot works on a blood cell.

n the *Star Trek* universe, the Borg is a species that has blended the biological with the mechanical to create superstrong, superintelligent creatures—but they are drones with no independent thought. Millions of nanobots course through the drones' bodies, controlling and empowering their every activity. Their mission, when not performing their assigned tasks, is to assimilate other organisms into the Borg collective by injecting them with nanobots as well. Only through great effort can the link between drone and collective be broken, giving freedom and free will to the drone.

Sounds scary, doesn't it? It's not completely far-fetched, though; after all, if a drug like heroin—biochemically much simpler than viruses—can change the brain chemistry of its addicts, forcing them to seek ever more heroin, why can't nanobots—which are basically mechanical viruses—turn their host organisms into viral automatons themselves? The technology won't be ready until long into the future; right now, work with nonbiological viruses is in its infancy. ■

DID YOU KNOW

One possible path for future genetic research is to construct DNA sequences on a computer and then create physical copies remotely, using a 3-D printer. (It is already possible to print meat grown from beef stem cells.)

THINK ON THIS ▶ **Should We Manipulate the Human Genome?**

"What is of the greatest value? What trait do you want to select for? . . . It may be that the best thing to have in 50 years is resistance to the Ebola virus. That may actually be much better . . . [and more] useful . . . than blue eyes and blond hair." —Bill Nye the Genetics Guy

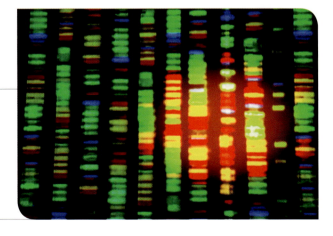

"Bending steel with his bare hands—I'm good with that. Flying in blue panty hose—I'm good with that. But taking energy from the Sun? No, no, not in a million years."

—DR. NEIL DEGRASSE TYSON, ASTRO-SUPERHERO-IST

CHAPTER TWO

Could Superman Survive a Black Hole?

What could be more fun than being able to climb up walls, vibrate through walls, or vaporize walls with a glance? Creatures with supernatural powers have appeared throughout history—but costumed superheroes have been part of pop culture for less than a century.

Their debut coincided with our first forays into modern science and technology—the expanding universe, quantum mechanics, transoceanic flights, and more. Superman, unsurprisingly, was an alien; Batman, a science-based detective; and Wonder Woman, an Amazonian who flew an invisible plane. As our knowledge progressed, so did our heroes—like the Fantastic Four, created by cosmic rays; the Hulk, victim of gamma radiation; Firestorm, the nuclear man; and cosmic superheroes Nova and Quasar.

The imaginative pseudoscience that comes from comic books inspires us to ask questions and seek answers while having fun doing it. And of course we wonder: Could we be superheroes too?

Humans can't travel through
a black hole, but could a superhero?

> *"As he was trying to run at the speed of light, he was going slower . . . It was a beautiful illustration of Einstein's principles."*
>
> —JAMES KAKALIOS, AUTHOR OF *THE PHYSICS OF SUPERHEROES*

Hulk may outsize Bruce Banner, but he'd weigh the same.

Cosmic Queries: Superheroes

Would Superman or Mr. Fantastic Survive Spaghettification?

Comic books play fast and loose with the laws of physics, but we don't. "If you can actually travel faster than light, you can just climb out of a black hole; nothing will stop you," says Neil. "However, the journey down to the singularity would, as sure as night follows day, spaghettify him . . . In principle, anybody who is stretchable [like Mr. Fantastic] might be immune to this."

▶ HOW CAN THE HULK GROW AND SHRINK?

"He can't just get bigger, unless he has the same mass," explains Neil. "And if he does, he's less dense in the state of Hulk than he is as Bruce Banner."

"So he'd be very kind of marshmallowy?" asks Chuck Nice.

Sure, says Neil. "Or he'd be like a beach ball."

▶ CAN MAGNETO CONTROL EARTH'S CORE?

"Yes, but just because something's made of metal doesn't mean Magneto can control it, because not all metal is magnetic," Neil clarifies. "Magneto could totally tear a new one in a star, because he can interact with the magnetic fields that control where those gases appear and what they do."

▶ WHAT KEEPS THE FLASH FROM BURNING UP?

The Flash gets his superspeed thanks to a fictional "speed force," which is fundamental to the universe, like gravity. An energy aura permeates and surrounds him as he runs—so he's protected from friction, which would otherwise set him on fire, and he doesn't get squashed if he runs into something at superspeed. ∎

Can We Increase Human Attributes, Like Captain America?

Captain America is human and superhuman at the same time. How does he do it? Could we do it, too? "We're all human at the end of the day," admits astro-muscle-ist Dr. Neil deGrasse Tyson. "Muscle size correlates with muscle strength. So you can have a Captain America who, if he had a good body, would turn heads in a fitness center, but the strength of those muscles is not such that he would have the strength exhibited by any genetic manipulation at all—and still have it be biological matter."

> *"Anything that any biological creature does that is special could, in theory, be incorporated into a human being."*
>
> —DR. LEE SILVER, MOLECULAR BIOLOGIST

There may be one caveat to Neil's explanation, though. If the Super-Soldier Serum changed the way Captain America's muscles are actually put together—that is, he's human, but slightly evolved beyond typical humans—then there could be more tensile strength per muscle fiber in his body, leading to greater strength even with the same size. And remember, he's superstrong for a man, but he's not Hulk-strong or anything. ∎

BACK TO BASICS

What Would NOT Happen If a Radioactive Spider Bit You?

The "spider powers" that Peter Parker received from the radioactive spider bite in the original comic books did not include, interestingly, his web. He actually invented both his web fluid and web shooters while he was in high school—either of which would have totally been a winning science-fair project. Although Spider-Man probably wields them better than anyone else, actually anyone with enough practice could use them.

Patriotic super-soldier Captain America attained peak human performance.

THINK ON THIS ▶ Would We Get Powers If We Used Our Whole Brain?

"The fact is, the idea that we use 10 percent of our brain was never true. It is a misquote from what actually was said by [a] neuroscientist a hundred years ago, who said, 'The brain is so complex we only know what 10 percent of it is used for.' So all this stuff with Professor X [telepathy] . . . and Lucy [telekinesis], this is a fiction." —Dr. Neil deGrasse Tyson, astro-debunk-icist

The Physics of Superheroes, the Sequel

Which Superpowers Will Technology Make Possible?

Technology already exists now that can do things superheroes do. Jet-powered wingpacks allow people to fly like an airplane. Backscatter x-ray imagers and active millimeter-wave scanners provide penetrating vision like Superman's. Maybe the most exciting superpower-like advances are in the areas of applied quantum mechanics: Quantum entanglement may someday allow teleportation; quantum computing could produce artificial superintelligence. Theoretical physicist Dr. Michio Kaku thinks that's the case as well: "If you could control the laws of the quantum, then you could in fact have . . . most of the superpowers of science fiction."

The biggest drawback for a lot of this technology is its portability. Can you carry that backscatter x-ray imager on your head as you fly through the air, for example? So these technologies will need to be miniaturized. ◾

NEIL TWEETS

Which Superhero Would Neil Like to Be?

"I like Batman. I can be Batman. And who doesn't like the gadgets? And who doesn't like the car? He's got the best car of any of them . . . You gotta like the car. As phallic as it is, nonetheless he's got cool things that the car can do. So the car is an extension of his utility belt in terms of its coolness factor. So for me, I'd have to be Batman."
—Dr. Neil deGrasse Tyson, astro-bat-icist

"I would have loved to have done Dr. Manhattan in Watchmen. However, I just worked with Billy [Crudup], who played Dr. Manhattan, and he was magnificent."
—LAURENCE FISHBURNE, ACTOR, ON WHICH SUPERHERO HE'D LIKE TO PLAY

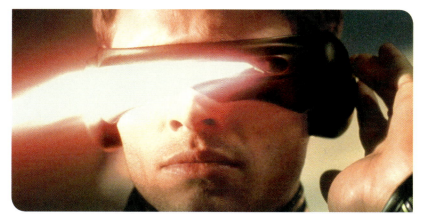

Cyclops from X-Men (2000) emits an optic blast through his visor.

The Physics of Superheroes

Could the Invisible Woman Really Disappear?

Invisible Woman in Fantastic Four: Rise of the Silver Surfer *(2007)*

n principle, making something invisible is easy: Just bend light from behind the object around it, so that whatever is behind it looks the same. We already know how that can be done in nature: A large concentration of mass—a black hole or a cluster of galaxies, for example, or even the sun in a weaker way—will act as a gravitational lens and bend light accordingly. Out in the universe, though, the lensing effect creates distortion in the shape and brightness of the objects behind it—a dead give- away for an unseen superhero.

> "[As a kid] I wanted to be Mighty Mouse. Because I wanted to save women when mean people were trying to harm them. And I wanted to sing opera while I was doing it."
>
> —DR. NEIL DEGRASSE TYSON, ASTRO-RODENT-ICIST

A more practical device has now been developed that might work better: a continuously multidirectional 3-D cloaking device. It can work only on small items now—a hand or a face, but not an entire person. Maybe someday, though, portable full-fledged invisibility systems will make the Invisible Woman's power seem like no big deal. (Her force fields, though, would still be quite formidable.) ◾

> "We analyzed [Wonder Woman's bracelets] from a materials science point of view . . . What are these bracelets made out of that can deflect bullets? . . . Cold rolled steel would probably be strong enough . . . Of course, her wrists don't snap, so she's excessively strong."
>
> —DR. JAMES KAKALIOS, PHYSICIST AND AUTHOR OF *THE PHYSICS OF SUPERHEROES*

THINK ON THIS ▸ How Does Athletic Wear Impact Superhero Movie Costumes?

"I always look to what sports figures do, and specifically what they do for the Olympics. That's the time when a lot of innovative technical fabrics are introduced for athletes . . . That's how it trickles down then to film. And, for instance, the costume designers for Spider-Man said, 'You know what? That looked really great, speed skaters, so let's appropriate that for the costume.'" —James Aguiar, fashion designer

A bungled nuclear physics experiment gives Dr. Manhattan of Watchmen (2009) superhuman powers.

Do Superheroes Teach Us Science?

"Mighty Mouse had a cape. Superman had a cape. So to me, it was clear: It was the cape that makes you fly. It was clearly the case . . . This was when I was in third grade or something."

—DR. NEIL DEGRASSE TYSON, ASTRO-CAPE-ICIST

Spider-Man's alter ego, Peter Parker (who shares a name, by the way, with a retired nuclear astrophysics professor at Yale), was a graduate teaching assistant for a semester or so. Seriously, though: In the same way that *Star Trek* communicators and tricorders have inspired mobile phones and medical imagers, superheroes make us think about what we haven't discovered yet.

Take, for example, Neil's desire for powers like that of the blue-skinned Dr. Manhattan: "In *Watchmen,* he has become a macroscopic quantum object, so where he can say, 'I can become this particle-wave duality myself. And I can, at will, become the wave and show up where the wave traveled in another spot and reassemble myself, and there I am.' Why can't we do that?"

Physicist Dr. James Kakalios has a simple explanation: "Because we don't have independent control over our quantum mechanical waves." ■

Could You Play Quidditch in Space?

Lots of magic users are superheroes—the Scarlet Witch of the Avengers and Zatanna of the Justice League come to mind—but they rarely are much fun. An airborne arena teeming with witches and wizards on broomsticks, though? Now that's a sport. Quidditch may be the greatest game ever invented in fiction and translated into reality—except that the version played in college club leagues today is strictly two-dimensional.

With only minor modifications, though, Chasers, Beaters, Keepers, and Seekers could ply their trade in space—with rocket packs instead of magical brooms. The scoring rings would have to be held in place somehow, presumably affixed to the edges of an orbiting stadium. The players' space suits had better be durable, though—it's all fun and games until somebody loses their airtight seal because a Bludger strike cracks a helmet or visor. ■

"The [Golden] Snitch, of course we presume, doesn't have broom powers but is actually aerodynamically supporting itself with flapping wings, like a hummingbird. Wings are useless in zero g . . . A bird on an airless planet is a brick. So they'd have to redesign the Snitch for that one."

—DR. NEIL DEGRASSE TYSON, ASTRO-QUIDDITCH-IST

BACK TO BASICS

Could We Replace a Skeleton With Metal?

From Wolverine of the X-Men, to Cyborg of the Teen Titans, to the Six Million Dollar Man and the Bionic Woman, exotic internal prosthetics have been the pseudotechnological source of many a superhero's powers. Neil points out what he sees is a flaw with this strategy: "Keep in mind that for the human body to work, you'd have to graft the tissue onto the metal. Our muscles connect to tendons that connect to ligaments that connect to your bones. It works biologically. In order to get foreign material in there, you're gonna have to attach it all some other way."

THINK ON THIS ▸ What If You Had Atomic Powers?

Dr. Solar, Man of the Atom, has been around in various incarnations since 1962. Among other things, he can transmute elements at will. So what could you do with power over the periodic table? Physicist Dr. Michio Kaku explains: "You get gold from nothing." And comedian Chuck Nice is all in: "OK. There you have it. Thank you very much, sir . . . I want that power."

Science Versus Superman

Faster than a speeding bullet! More powerful than a locomotive!
Able to leap tall buildings in a single bound! Just how super is Superman,
subjected to the scrutiny of science? Here are Neil's answers
to the essential questions about Superman.

||||||||

▶ WHAT'S WITH THE YELLOW SUN THING?

"We know the difference between a yellow star and a red star . . . It's just light. So if it's light that gave Superman his powers and if it's red light that took them away, then all you'd need to do is shine red light on Superman and he'd be a crying mess."

||||||||

▲ DOES HE KNOW ASTRONOMY?

"Superman wanted to find his home planet, Krypton, because the light signal of Krypton, [once it was] destroyed, would be coming to us about now. And he came to the planetarium. So I'm director of the planetarium . . . So I'm there with Superman."

||||||||

◀ SHOULDN'T HE BE RADIOACTIVE?

"You can't store [solar] energy in an unlimited way without getting hot—as in radioactive. It's just how nature manifests itself. He'd be very hot, and you'd know he was coming from a long way."

||||||||

◀ HOW DOES HE FLY?

"What you have to rely on is his muscular strength from a planet that may have a very high surface gravity . . . If you're that strong, it looks like you're flying when you're just leaping."

◄ COULD HE AND LOIS LANE PROCREATE?

"I would say, Superman looks so humanoid that there has got to be sufficient overlap there to try [for] a cross-species baby . . . Watch out: The baby could kick out, like in *Alien*."

▲ WHO WOULD WIN: SUPERMAN OR THE *ENTERPRISE*?

"I have no doubt that Superman can take the starship *Enterprise* and the entire crew . . . If Superman goes to the tail of the *Enterprise* and punches it or swings it around lasso-style, that's the end."

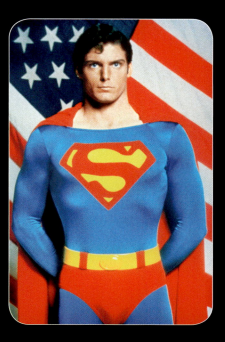

▲ HOW IS HIS VISION "X-RAY"?

"If he has x-rays . . . it's not going to see the color of her panties. The x-rays would go right

▲ WHAT'S DEADLIER: HIS GAS OR HIS WIND?

NEIL: There's physics in everything. That's all I'm saying.

"So what's science fiction? Science fiction serves a good purpose. Because it sets those goals, those benchmarks out in nowhere. And we work toward that. And then—now—we surpass that."

—GEORGE TAKEI, ACTOR WHO PLAYED SULU ON *STAR TREK*

CHAPTER THREE

Why Don't We Have Flying Cars Yet?

By today, said the science-fiction writers of the 20th century, we all would be zooming around above our cities. Robots would be catering to our every need. And we would be living on other planets and beyond.

We should thank our lucky stars they were wrong. Otherwise, by now we'd be driving desperately around a destroyed society or enslaved by mechanical masters or stranded on distant outposts, yearning for the green shores of our home world. At least we have smartphones, solar panels, search engines, and space stations. It could be a lot worse.

Science fiction points out possibilities for us to ponder. It affects how we think, where we live, when we laugh, and even what we wear. It may try to be realistic, or not. It just has to be believable enough for us to see beyond the fantasy and look to the core of who and what we are. Science fiction allows us to examine ourselves from the outside. Let's peek now at what's inside!

We haven't reached an age of flying cars—but are we close?

Science fiction can be about the future, the past, or both.

What Is Science Fiction?

According to the Gunn Center for the Study of Science Fiction at the University of Kansas, sci-fi is fundamentally "the literature of the human species encountering change." It can inspire, entertain, and warn us about what was, is, and may be.

Science fiction is usually about the future, but it can be about the past—especially in time-travel or alternate-reality stories. And in some cases—like H. G. Wells's novel *The Time Machine* or the 2006 movie *The Fountain*—it can be about both.

▶ WHERE ARE THE FLYING CARS?

"Are you sure you want a flying car?" asks Elon Musk, founder of Tesla Motors and SpaceX. "If there are flying cars, then, well, obviously you've added this additional dimension where now a car could potentially fall on your head."

▶ HOW REAL IS THE SCIENCE IN SCI-FI?

Sometimes, it's right on. Other times, not so much—like Han Solo making the Kessel Run, says Neil. "[A parsec] is a unit of distance, and there he is boasting about the speed of the Millennium Falcon in 12 parsecs. So this is a completely scientifically illiterate statement. Later on, there'd be this revisionist discussion . . . 'Oh, no, what he meant was the Millennium Falcon went past a warp in space-time, making the distance shorter.' " ■

"We had this amazing device that was attached to our hip. And we would walk around with it all the time. And whenever we wanted to talk to someone, we'd rip it off, flip it open, and start talking. At that time, it was an astounding piece of technology. Today we've gone way past that."

—GEORGE TAKEI, ACTOR WHO PLAYED SULU, DESCRIBING *STAR TREK*'S COMMUNICATORS

Did Dr. McCoy's Tricorder Lead to the MRI?

Wil Wheaton, the actor who played Wesley Crusher on *Star Trek: The Next Generation,* offers a bit of trivia: "You know, the guy that invented the MRI . . . watched the original Star Trek—and he watched Dr. McCoy sort of scan around on a thing, and he thought, We should do that; there should be a way that we can see inside people's bodies without having to cut them open. "

MRI, or magnetic resonance imaging, is a great diagnostic tool—but it's loud, bulky, expensive, and time-consuming. So what will the MRI lead to? How about a real tricorder? "[The] Tricorder X Prize is challenging teams to build a device that an average consumer can use, that you comp on, that you can talk to, that you can do a finger blood prick to, and it can diagnose you better than a team of board-certified doctors," says Peter Diamandis, founder of the X Prize. ∎

A replica of Dr. McCoy's tricorder

NEIL TWEETS

What Was the Genius of *Star Trek* (and *The Twilight Zone*)?

"[*Star Trek*] was like nothing that came before," says Neil. "Yes, there was science fiction. The difference was, this told stories that really should have been told in real-Earth situations, but no one would have allowed that to happen, because they were offensive, or they probed our odd social mores. So you could transpose it into space and have these stories still be told. *The Twilight Zone* had an equivalent landscape in which to conduct the storytelling."

"It seems to me that the preservation of life and extension of life would be a natural consideration for cyborg development. I mean, we have artificial hearts, artificial other components . . . I think there are people alive today that will see things like this."
—STEPHEN GOREVAN, ROBOTICIST AND SPACE SCIENTIST

THINK ON THIS ▶ Why Was Geordi's VISOR on *Star Trek* So Clunky?

"I love the VISOR," says LeVar Burton, the actor who played Geordi La Forge on *Star Trek: The Next Generation.* "But I always wondered: If our technology was so damn sophisticated, why couldn't we put the technology in something a lot smaller than the VISOR?" Technology progresses in science fiction, too: By the eighth *Star Trek* movie, Geordi was seeing through artificial eyes.

In the premier episode of Star Trek *(1966) the* Enterprise *crew leave the ship via transporter.*

What Would Actually Happen in a Transporter?

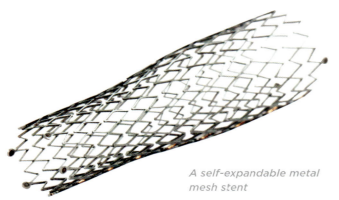

A self-expandable metal mesh stent

I n the *Star Trek* transporters, your atoms and their information were recorded, disassembled, sent in an energy beam, and then reassembled at a distant location. "The problem with that is that [if] you convert your mass into energy, that would be an explosion larger than every nuclear bomb on the planet. That would be bad," explains Dr. Phil Plait, astrophysicist and author of *Bad Astronomy: Misconceptions and Misuses Revealed.*

The complexity of transporting goes to the quantum level, too. We all have at least a thousand trillion trillion atoms in our bodies, so scanning and recording even one person's most basic atomic information would far exceed all the current computer storage capacity on Earth.

"How about transporting a stent into your artery so you don't have to do any surgery?" asks Dr. Charles Liu, astrophysicist.

Maybe it's best to start simple, think small, and learn to transport useful but inanimate objects for now. ▪

Could the Death Star Really Blow Up a Planet?

Zapping a planet-size object *Star Wars*–style with a single powerful beam of energy looks great, but it doesn't really work. "It turns out . . . it's almost impossible," admits astrophysicist Dr. Phil Plait. "You could take something the size of Mars and hit the Earth with it, and not completely destroy it. So to completely destroy a planet . . . it takes a huge amount of energy . . . When you calculate it, it's like more energy than the sun puts out."

The more recent *Star Trek* version of planetary destruction—collapsing the planet with its own gravity into a singularity—makes more sense energy-wise and mass-wise—but it's got reality-check problems of its own, like the utter impossibility of so-called "red matter."

Astro-kablooey-ist Dr. Neil deGrasse Tyson explains: "So you calculate the binding energy of the planet. And if you pump more energy than the binding energy into the planet, it'll explode in the way the Death Star destroys a planet . . . You can destabilize a planet if you make Swiss cheese out of its innards—that's clear. But I like the *Star Wars* style of destruction. Just a good old-fashioned planetary explosion." ■

NEIL TWEETS

Neil and Brian Cox's Lightsaber Twitter War

@NEILTYSON: If LightSabers are made of light, they would just pass through one another.

@PROFBRIANCOX: Not if the photons are sufficiently high energy.

@NEILTYSON: We could trap the escaping particles within one of the swords.

@PROFBRIANCOX: The charged particles produced could be magnetically contained.

"Maybe the name is just wrong. Maybe they just call it a lightsaber. That doesn't mean it's made of light."

—DR. PHIL PLAIT, ASTROPHYSICIST AND AUTHOR OF *BAD ASTRONOMY: MISCONCEPTIONS AND MISUSES REVEALED*

THINK ON THIS ▶ How Would a Lightsaber Actually Work?

DR. BRIAN COX: At very high energies—very high-energy collisions—there is a probability that photons will kick off each other, bounce off each other.

DR. PHIL PLAIT: Maybe the name is just wrong. Maybe they just call it a lightsaber. That doesn't mean it's made of light, if it's a force field that comes out and it's filled with a plasma or something like that.

> *"Well, the* Armageddon *movie had a few flaws. Surprise, surprise. Shocking! But the thing they got mostly right is that if a very large comet were to come in, it would certainly cause a lot of destruction."*
>
> —DR. AMY MAINZER, ASTROPHYSICIST

Bruce Willis in Armageddon *(1998)*

DID YOU KNOW

In *Armageddon,* an asteroid the size of Texas was discovered just a few weeks before it would hit Earth. There are only four asteroids that big in the whole solar system— and astronomers had found them all by the year 1849.

Was *Armageddon* Accurate? (And Other Sci-Fi Movie Reviews)

In what ways *wasn't* it inaccurate? If rogue comets or other space objects were to create lots of gravitational interactions in the outer solar system, asteroid orbits could be disrupted—and Earth could be pelted with small solar-system objects. Something like that happened once; it's called the Late Heavy Bombardment. Lucky for us, though, it ended around 3.8 billion years ago.

▶ WHY DID THE *ENTERPRISE* RISE FROM TITAN?

Planetary scientist Dr. Carolyn Porco tells this story about the 2009 *Star Trek* movie: "I told [director J. J. Abrams], 'Have it come out of warp drive in Titan's atmosphere . . . and have it rise out, submarine-style, out of the clouds, with Saturn and the rings in the background. It'll be a knockout scene.'"

▶ WHAT WOULD FROZEN OXYGEN FROM SPACE DO?

In the late 1970s TV series *Buck Rogers in the 25th Century,* Earth authorities bring a huge chunk of frozen oxygen to Earth in order to replenish the planet's dwindling supply. Really? "What will happen is, fires that are ignited will not burn out as easily as they might today, because oxygen feeds fires," says astro-pyro-cist Dr. Neil deGrasse Tyson. "You would burn forests like it was nobody's business."

▶ HOW SCIENTIFIC ARE *DOCTOR WHO*'S TOOLS?

Not scientific at all. That doesn't mean their fictional explanations aren't fun anyway, says Dr. Phil Plait, astrophysicist. "I was like, 'That's awesome. That sounds like it makes sense.' And then you go, 'Wait a minute' . . . And that's what I love about *Doctor Who.* They explain things, but you don't really gain any knowledge when you walk away." ■

"Forget your other organs—it's your brain you care about. You get a machine that lives forever, put your brain in the machine."

—DR. NEIL DEGRASSE TYSON, ASTRO-MAD SCIENTIST

StarTalk Live!: I, Robot (Part 2)

Will We Ever Be Able to Put a Human Brain in a Robot?

Bill Clinton, the 42nd president of the United States, talks about the future of robotics and medicine: "I keep meeting all these friends of mine that have two new hips and knees and all this . . . It looks like we're going to have replacement parts for just about everything. But if they replace your brain, would you still be you? Will there be the equivalent of like a SIM card or some sort of hard drive that takes everything out of your brain if you get a new one, and puts it back in the new brain?"

Well, Mr. President, the complexity of the brain far exceeds the storage capacity of any computer disk or card we have today. And the brain's synaptic structure changes so quickly that any digital backup of you probably wouldn't be you anymore in a matter of minutes—or even seconds. So go with a brain transplant rather than a brain backup. ◼

TOUR GUIDE

Do Drones Follow Asimov's Laws of Robotics?

In Isaac Asimov's Robot novels, it is posited that humans have the wisdom to program robots not to harm humans, disobey humans, or destroy themselves, except under certain limiting conditions. The level of autonomous robotic ability that current military drones have—to plot a course to home base, for example, if a remote-control signal is lost—would not rise to the level of robotic intelligence expressed in Asimov's noble laws. If drones were to become that intelligent, who knows what they'll be programmed—or, on their own, decide—to do.

"Yeah, do you want to be buried, cremated, or uploaded? You get a choice."

—JASON SUDEIKIS, COMEDIAN

THINK ON THIS ▸ **Do Artificial Beings Have Human Rights?**

Brent Spiner, who played the android Data in *Star Trek: The Next Generation,* describes an episode that meant a lot to him: "[It] was an episode called 'The Measure of a Man,' in which my character was on trial, basically to decide whether or not he . . . was a sentient being . . . and whether, if he was not sentient, were we creating a race of slaves . . . or did he have his own right to his own existence?"

Sci-Fi Fashion: Forward or Retro?

"When you look at a futuristic movie or a science-fiction film that some designer or costumer designer spent beaucoup time designing, it . . . inevitably ends up looking like the time we're living in. So can you really design the future?"—James Aguiar, fashion designer

IIII
◀ *STAR TREK* (1966)

Go-go boots, miniskirts, and bell-bottoms were the cutting edge of '60s fashion.

IIIIIIII
◀ *MAD MAX: FURY ROAD* (2015)

This version of postapocalyptic Australia featured the millennial version of punk couture—and won the Oscar for best costume design.

IIIIIII
▲ *INTERSTELLAR* (2014)

To reflect a not-so-distant future, *Interstellar* relied on dress shirts and tees similar to today's fashion.

IIIIIII
▶ *2001: A SPACE ODYSSEY* (1968)

The color-blocked and square cuts of the steward-esses and astronauts in the film created an oddly time-less futurism in the movie.

||||||||
◀ *STAR WARS* [1977]

Loose, flowing clothing invoked exotic magic, contrasted with the evil-looking white armor of the Imperial stormtroopers.

||||||||
◀ *THE MARTIAN* [2015]

Beneath the comfy-looking, formfitting space suits were body-hugging—but practical—unisex T-shirts and fleece.

||||||||
▲ *DUNE* [1984]

Clothing worn by a technology-fearing feudal society 10 millennia in the future mixed medieval, military, and 1980s fashion sense.

||||||||
◀ *PLANET OF THE APES* [1968]

The apes' outfits were a metaphor: Beneath veneers of cloth, we're all just animals within.

||||||||
▲ *AVATAR* [2009]

Lots of bare skin and long braided locks produced a primal back-to-nature—dare we say hipster—look.

Artwork depicts events within specific cells of living tissue.

Is Brainwashing Real?

Hypnotists are often shown with a swinging pocket watch.

Human brains—and thus people's behaviors and personalities—are malleable, as evidenced by radical terrorists and fanatic cults. Can that kind of control now be exerted with machines?

Neural prosthetics, such as cochlear implants for hearing or spinal cord stimulators for pain relief, are already widely used in human medicine. Amazingly, scientists using neural prosthetics can now control some rudimentary animal behaviors. Flying insects with electrodes in their brains can be commanded, for example, to turn left or right with a computer or joystick.

Optogenetics is another part of neuroscience, where parts of the brain are both scanned and stimulated using light signals. Optogenetic devices are aimed at communication—interpreting the brain's electrical signals and then sending messages back with light. Could memories and instructions be implanted as well? The implications for medicine—and possibly espionage—are staggering. ∎

Cosmic Queries: A Stellar Sampling

Can HAARP Manipulate the Weather?

From 1993 to 2015 the U.S. Air Force constructed and ran the High Frequency Active Auroral Research Program (HAARP), probing the ionosphere to see if radio signals could be enhanced in Earth's upper atmosphere to improve wireless communication or surveillance.

Rumors began to spread. Conspiracy theorists—including Venezuela's president at the time, Hugo Chávez—claimed HAARP was being used for nefarious tasks: shooting down an airplane, destroying the space shuttle *Columbia,* controlling minds, spreading disease, and creating storms, floods, earthquakes, and global warming. Even though it seems clear that these claims are false, a good scientist is willing to consider the evidence.

Here's Neil's reaction: "There are people who are sure that the government is stockpiling aliens and controlling everything about anything we would ever think about, and they clearly have never worked for the government . . . I have not been convinced by any of the reports to suggest that experiments in the upper atmosphere, physics experiments, are having any effect on our weather whatsoever." ■

BACK TO BASICS

How High Could a Kite Fly?

"The higher the kite goes, the more string is dangling beneath it. You reach a point where the weight of the string rivals the weight and updraft buoyancy of the kite . . . If you had a really huge kite, you could take it up to the stratosphere. The problem is, in the stratosphere, wind speeds are several hundred miles an hour." —Dr. Neil deGrasse Tyson, astro—Mary Poppins—ist

DID YOU KNOW

Research is being done on the possibility of flying windmill-like kites thousands of feet in the air to generate electricity from high-speed winds.

THINK ON THIS ▶ Does the Archimedes Mirror Attack Work?

According to legend, the ancient Greek inventor Archimedes set fire to invading Roman ships using bronze shields to focus sunlight. In 2005 Dr. David Wallace at MIT was able to simulate such an attack under controlled conditions. But when he and the hosts of TV's *MythBusters* tried it with 500 high school students on shore pointing big mirrors at a boat in the water, it didn't work.

An imagined future city on the coast

StarTalk Live!: Building the Future

What Will Cities of the Future Be Like?

The gleaming spires and crisp lines of futuristic space cities have long been a mainstay of science fiction and its high-tech visionaries. They sometimes seem to be here already; just take a look from a distance at the skylines of New York, Shanghai, London, or Dubai. But dystopic visions of decaying urban wastelands have also been shown, perhaps most famously in the 1982 film *Blade Runner.* Can technology be used to stave off the ugly and send in the beautiful? If so, when?

"With a flying car, you're talking about going 3-D . . . But then you go to the street and suddenly you're two-dimensional . . . Have more car tunnels, [and you] alleviate congestion completely."

—ELON MUSK, TECH ENTREPRENEUR

"We have smart materials, adaptive structures," begins futurist Melissa Sterry's prediction. "We have entire buildings that can move. And we have the sensors, the information systems to inform those processes . . . The built environment is on the cusp of acknowledging, 'My God, we've actually got to get with the times. We've got to take advantage of these new opportunities' . . . We will have very smart cities by the year 2020 . . . The kind of bionic city I'm talking about, I'm giving a timeline of between 2040 and 2050." ∎

Homer appeared as a robot in The Simpsons' "Days of Future Future" episode (2014).

Do Science, Sci-Fi, and Comedy Mix?

On Family Guy, *writer Seth MacFarlane often puts character Stewie in sci-fi situations.*

Many of us, sadly, know science only from humorless teachers and their classes. But in the right hands, science and comedy can be as tight as a tick on a tauntaun.

"I was always a big science-fiction fan, so I whenever possible jumped at the opportunity to have Stewie veer into that world," says *Family Guy* creator Seth MacFarlane. "Because certainly animation fans tend to be fans of science fiction and have an interest in science."

A few years back, psychologist Dr. Nina Strohminger published a scientific study with an earthshaking conclusion: "Farts make everything funnier." From speculations of Superman's destructive flatulence to discussions of astronauts shooting around the ISS, astro-silent-but-deadly-ist Dr. Neil deGrasse Tyson has compiled plenty of evidence to support Nina's claim. "But, just to be precise, that's exactly how the Shuttle lifts. It expels gas out of one end, and the physical shuttle recoils in the other." ∎

If in Space, No One Can Hear You Scream . . .

Why did we hear explosions in *Gravity*? Movies don't have to be scientifically accurate to be entertaining. *Star Wars*, one of the great sci-fi movie franchises of all time, is fundamentally untrue to the laws of physics, but we like it anyway. How about closer-to-Earth movies, like *The Martian*? Here are a few comments from Neil, many of them tweeted right after he saw the movie.

‖‖‖‖
◀ **GRAVITATIONAL HAIRSTYLING**

@neiltyson: "Mysteries of #Gravity: Why [Sandra] Bullock's hair, in otherwise convincing zero-G scenes, did not float freely on her head." Must be using really good astronaut mousse.

‖‖‖‖
▶ **BETTER TOGETHER**

@neiltyson: "Mysteries of #Gravity: When [George] Clooney releases Bullock's tether, he drifts away. In zero-G a single tug brings them together." But then where's the drama in that?

‖‖‖‖
◀ **WHAT'S IN A TITLE?**

@neiltyson: "Mysteries of #Gravity: Why Bullock, a medical Doctor, is servicing the Hubble Space Telescope."
@neiltyson: "Mysteries of #Gravity: Astronaut Clooney informs medical doctor Bullock what happens medically during oxygen deprivation."

◀ SAY IT AIN'T SO!

"Yes, absolutely, the sandstorm [in *The Martian*] is inaccurate . . . I just didn't care. I wanted a good reason to strand him there, and at the time I wrote it most people didn't know that." —Andy Weir, author of *The Martian*

|||||||

▲ MARTIAN GARDENING

The Martian's greenhouse is full of potatoes? As long as there is enough water and grow lights, maybe. Essential soil nutrients for potato growth, though, are probably lacking, even with fertilizer.

|||||||

▲ SPACE TRAVELERS

The rotating sections of *The Martian*'s spacecraft can indeed create acceleration that mimics gravity on a large spaceship. We still need some engineering innovations to make it possible, though.

|||||||

▲ STAR POWER

@neiltyson: "In @StarWars #TheForceAwakens, if you were to suck all of a star's energy into your planet, your planet would vaporize." Bye-bye, Starkiller.

|||||||

◀ SPACE VOLUME

@neiltyson: "In @StarWars #TheForceAwakens, the TIE fighters made exactly the same sound in the vacuum of space as in planetary atmospheres." Same goes for the X-wings and the Millennium Falcon.

"It's a very strong desire for humans to believe in things in the sky . . . And it's related to the whole concept of humans wanting to find meaning and purpose . . . UFOs fit this bill perfectly."

—MAJ. JAMES MCGAHA, ASTRONOMER AND FORMER USAF PILOT

CHAPTER FOUR

Could Bigfoot Be a Space Alien?

n a story arc of the classic 1970s show *The Six Million Dollar Man,* Bigfoot, also known as Sasquatch, is revealed to be a space alien—misunderstood and mistreated, but overall a really nice guy.

So that was a work of fiction. It wasn't real! Tell that to all the people who are sure they've seen a giant manlike creature wandering the woods with them. Will they believe you? How about the people who have seen UFOs and big-eyed extraterrestrials? The tricks our brains can play about what we think we've experienced can be complex and astonishing—and those are the same tricks other people exploit to befuddle us. Will their efforts defraud us or delight us? If we know what the tricks are, we can make a choice.

With the scientific confirmation of thousands of planets beyond our solar system and the near certainty of finding more, can finding aliens be far behind? Imagine what'll happen if we succeed.

A friendly meeting of Bigfoot and an extraterrestrial visitor

DID YOU KNOW

Stephen Colbert loves having scientists on his show. "Add comedy to the fascination of their subject . . . that's a honey ball that might just make people question the world around you."

The Science of Illusion With Penn & Teller

Is the Human Mind Programmed to Believe?

"We are wired to find meaning. We try to connect cause and effect all the time, whether cause and effect are connected in the real world or not. This is what happens in a magic show: . . . It's the magic wand [that] makes the rabbit disappear. In reality, the cause is something very different. You go to a psychic and they give you something that looks like very tangible evidence—and you may not want to stop and think about it, because you're missing that person and you're in a vulnerable position, so you're not thinking clearly to begin with."
—Dr. Susana Martinez-Conde, neuroscientist ■

A crystal ball inverts the image behind it—but will it tell the future?

BACK TO BASICS

Would You Like to See That Again?

"We are going to trick you. We are going to use your mind against you. And we're going to do that in the spirit of sharing and pleasantness. That's a difficult social contract, and other magicians don't want to do that . . . But if I come to you and say, 'You know, we don't have any powers. But you know, Neil, there's this way of handling books and talking to people that makes it look like I can read what's in your mind. Isn't that kind of nutty? Let's give it a try.' And all of a sudden, we're on the same side."
—Penn Jillette, magician

"We're all lying to ourselves and to each other all the time. Magicians just do it better . . . At least in part, what we're experiencing the vast majority of the time is an illusion."
—DR. SUSANA MARTINEZ-CONDE, NEUROSCIENTIST

A Conversation With Dan Aykroyd [Part 2]

Are Crystal Skulls Evidence of Aliens and Gods?

I n the late 1800s, crystal skulls started appearing in souvenir shops in Mexico and in museums worldwide. They were said to be of pre-Columbian origin, recovered from Aztec or Maya sources, and soon enough, storytellers started talking about the skulls as "gifts" from the "star children." But up to the present day, no archaeological excavation has ever recovered a crystal skull. The idea makes for fun fake-archaeology movies, though.

▶ CAN YOU PROVE I DIDN'T SEE BIGFOOT?

Science is based on obtaining evidence that can be analyzed to confirm or refute a claim. The burden of proof is on the party making the claim; you have to do much more than just declare it to be true. So even if someone can't prove you didn't see Bigfoot, it doesn't mean you did.

▶ WHY MIGHT BIGFOOT HAVE COME TO EARTH?

"It would be kind of cool if Bigfoot—if some powerful alien civilization—said, "Earth? There's a good place to send our criminals. Let's send Bigfoot there," Neil says. But what crime did Bigfoot commit?

"Stole a candy bar," suggests comedian Leighann Lord.

▶ WHAT WAS *MYTHBUSTERS*' WEIRDEST CONFIRMED MYTH?

According to the hosts, Jamie Hyneman and Adam Savage, the weirdest myth they ever confirmed wasn't paranormal at all. It was about whether elephants really are afraid of mice. "So the elephant comes out on cue, and darned if it didn't come screeching to a stop once the mouse came out." ■

"It's not proof that [ghosts or Bigfoot] don't exist, it's just proof that you couldn't find it."
—JAMIE HYNEMAN, MYTH BUSTER

This crystal skull is most likely a hoax.

"It turns out, if you're blind, the Ouija board doesn't work very well for you . . . And it turns out, if you don't know how to spell well, it actually misspells the words . . . So unless your spelling profile exactly matches that of the dead person whose spirit you're channeling, this is not working."
—DR. NEIL DEGRASSE TYSON, ASTRO-SKEPTIC-IST

The Science of Illusion With Penn & Teller

Are Eyewitness Accounts of Miracles Lies or Illusions?

"Conspiracy theories are lazy."
—BILL NYE THE CONSPIRACY GUY

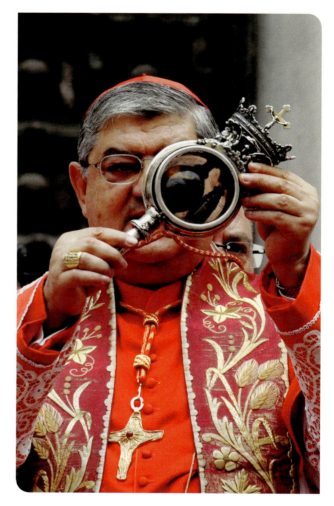

Research has clearly shown that, statistically speaking, eyewitness accounts of anything by anyone are pretty unreliable—and often, that's all the evidence we have to go on. You may not be lying about what you think you saw—but what did you really see?

"When it comes to eyewitness testimony, on which so much of our legal system is based, it's the most unreliable testimony that there can be," says comedian Chuck Nice. "And normally, the testimony ends up with somebody who looks like me committed the crime. Which is a little rough."

Neuroscientist Dr. Susana Martinez-Conde agrees, and has an explanation for why this happens: "When you give a testimony about a crime, you have witnessed a situation in which, almost by definition, there was high emotional content. You might have been angry or scared, or a lot of things were happening, and we know . . . that you cannot pay attention when you're experiencing a strong emotion." ∎

Cardinal Crescenzio Sepe displays the annual miracle of the blood of St. Januarius in Naples.

"Eyewitness testimony [is] the most unreliable testimony that there can be . . . The problem is, what you're remembering is very real to you. Just like when you saw the illusion, it was very real to you."
—CHUCK NICE, COMEDIAN

THINK ON THIS ▶ Can Conspiracy Theories Slow Progress?

Believing a conspiracy theory is essentially a decision to be willfully ignorant of reality and avoid thinking critically. And ignorance has often slowed scientific, medical, or social progress. If we believe vaccination is a conspiracy, for example, and thus don't vaccinate our kids, many more children could sicken and die. If enough people believe the moon landing was faked, might politicians defund the space program? That's extreme, but not impossible.

An imagined sinkhole in the Bermuda Triangle

YouTube: Neil deGrasse Tyson Explains the Bermuda Triangle

What's Going On in the Bermuda Triangle?

The pulp science-fiction magazine Amazing Stories *launched in 1926.*

The stretch of ocean between Bermuda, Puerto Rico, and Florida has been traveled extensively since the time of Christopher Columbus. It's a zone where warm and cold conditions can intersect, leading to storms and rough waters. Atlantic hurricanes track through all the time.

Even in modern times, human choices and bad seas can lead to tragedy. In 2015 an experienced freighter captain steered his ship, *El Faro,* from Florida toward Puerto Rico just as the powerful Category 4 Hurricane Joaquin was moving through the area. The ship began taking on water and then lost radio contact. *El Faro* was never seen again.

Statistically, however, there are no more shipwrecks or plane crashes in that region than would be expected, given the amount of traffic and weather there. Or as Neil puts it: "You ever notice there are never any missing trains? Trains never disappear." ∎

The Out-of-This-World UFO Show

What Is the Government Hiding in Area 51?

Maj. James McGaha, astronomer and former pilot, had top-secret clearance during much of his career with the U.S. Air Force and worked in Area 51—also known as Dreamland or, more officially, Homey Airport and Groom Lake, Nevada. All work done there is classified—so hush-hush, in fact, that the U.S. government didn't even officially acknowledge its existence until 2013. Stealth aircraft, such as the F-117 and B-2, were probably tested here, but anyone who knows for sure isn't telling. "This goes right to the heart of conspiracy theories," says Maj. McGaha. "UFOs are wrapped in conspiracy theories."

The secrecy surrounding Area 51 is perfect fodder for wild stories about alien spacecraft, alien technology, even aliens themselves at the site. After all, any flying object coming in or out of there is secret—and thus, by definition, an unidentified flying object. The only site that might be as commonly fictionalized in UFO lore is Roswell, New Mexico—the purported site of a 1947 UFO crash. ■

TOUR GUIDE

Why Search for Aliens When They're Already Here?

From *My Favorite Martian* to *The X-Files*, television shows have long posited in no uncertain terms that extraterrestrials are among us right now. Here is what Dr. Seth Shostak, senior astronomer at the SETI Institute, has to say about that: "I get at least five phone calls and emails a day from people who are having difficulties with aliens in their personal lives. They send me photos; they send me videos of UFOs. They often think that we're doing the wrong experiment, trying to eavesdrop on alien broadcasts, because after all, they—like one-third of the population of this country—believe that the aliens are here."

A mailbox near Area 51 is a landmark on the Extraterrestrial Highway.

"Conspiracy theories are very attractive, because they can explain complex social structures and problems."

—MAJ. JAMES MCGAHA, ASTRONOMER AND FORMER USAF PILOT

Is That a UFO or a Cloud?

There are thousands, maybe millions, of UFO sightings every year. Many of them are easily explained as helicopters, planes, or human spacecraft. (Yes, you can see the International Space Station from Earth, and it looks really funky.) Almost all of the rest of them are natural phenomena like bright planets (especially Venus), lightning and other weather events, meteors, or cloud formations. Why do people so readily interpret what they see as alien forms? Dr. Neil deGrasse Tyson, astro-cognitive-ist, has an answer: "What happens is, when you see something unfamiliar in the sky, the brain tries to understand it and come to terms with it.

"I think they've actually studied Earth and have concluded that there's no sign of intelligent life. What hubris it is to suggest that they would want to visit us in the first place."

—DR. NEIL DEGRASSE TYSON, ASTRO-HUMILITY-IST

It'll go overboard in doing so, and so what it doesn't understand, it connects the dots, it invents missing pieces, and so, when you try to judge what an object is, or you try to judge how big it is, your brain fills in the gaps."

One particular kind of lens-shaped cloud is often mistaken for a UFO—or, even better, interpreted as an artificially generated cloud designed to disguise a UFO. Standing lenticular clouds occur naturally, low in the atmosphere, usually hovering over a natural land obstacle like a mountain or ridgeline. They're naturally beautiful—and can look just like a fluffy alien flying saucer. ∎

Cosmic Queries: UFOs

Why Do Flying Saucers Spin?

Flying saucers are sometimes depicted—or purportedly observed—to be spinning as they fly. It's logical: Spinning creates acceleration toward an object's outer edge, which could mimic gravity as long as the passengers stood with their feet against the edge. But if they stood upright or sat facing forward, there'd be a lot of flying furniture and some dizzy aliens.

Only a fraction of saucer-shaped UFO reports over the decades actually describe the saucers as rotating. Hollywood depictions aren't consistent either; for example, compare *The Day the Earth Stood Still* (1951) with *Earth vs. the Flying Saucers* (1956). As for the ones that do appear to spin, it could be an optical illusion or just a thin outer shell that's spinning while the interior remains still. ■

EUGENE MIRMAN: *Your problem is that the whole saucer spins, and that's what a fool would build.*

NEIL: *Those would be stupid aliens. If they managed to do that, it would be violating very well-tested laws of physics.*

DID YOU KNOW
In the 1950s a Canadian company was contracted to develop a saucer-shaped aircraft for the U.S. Air Force. Project 1794 ended without success, but it did produce some valuable aviation technology.

TOUR GUIDE

Haven't We Already Been Visited?

Maybe aliens already came here—not to colonize or conquer but just to stop by, take a look, or get some work done, and then take off again.

If that's so, as the physicist Enrico Fermi asked, where are they? According to the Fermi paradox, there should be all kinds of extraterrestrial activity, and our astronomical searches should have found it by now—unless the aliens went to a lot of trouble to disguise their activities. But what would be the point?

THINK ON THIS ▶ Why Would Aliens Want to Come Here Anyhow?

ALAN ALDA, ACTOR: "These aliens are not going to come here, because the first thing they saw were our television programs . . . The first thing they heard was Orson Welles's radio show about the 'War of the Worlds' . . . They think the place is overrun by tourists already."

JOHN HODGMAN, COMEDIAN: "Think of the amount of energy and resources it would take to reach this planet. What would be gained?"

What Would Meeting Aliens Be Like?

H. G. Wells's The War of the Worlds *described fictional Martian attacks.*

From horrible destruction in movies like *Independence Day* to benevolent greetings in *Star Trek: First Contact,* alien meetings with humans have long been imagined based on either fearful or wishful thinking.

One voice on the side of caution comes from physicist Dr. Stephen Hawking, who said it might not go well for us humans—we may, he mused, wind up in the same position as the Native American population when Columbus arrived. "The fear factor that Stephen Hawking is expressing is more a fear of how he knows we would treat other people than it is from any actual knowledge of how actual aliens would treat us," Neil says.

If Hawking's concerns are legitimate, then humans may need to take precautions to make sure we're never detected. Maybe we'd need to encrypt our electronic transmissions and wireless communications—so that when they go into space, they seem like random radio noise. This, suggests former NSA contractor Edward Snowden, might be why we're missing out on alien communications. ◾

DID YOU KNOW

A new scientific survey for extraterrestrial intelligence called Breakthrough Listen will, over the next decade, search one million stars in our Milky Way and one hundred galaxies beyond it for signs of alien life.

THINK ON THIS ▶ **Was the "Wow!" Signal an Alien Message?**

In the 1970s a researcher working on the search for extraterrestrial intelligence (SETI) noticed a brief radio signal that looked so much like a possible coherent message that he circled it on the printout and wrote, "Wow!" next to it. The signal has never been heard again. Was it random noise? Have the aliens stopped transmitting—or started disguising their broadcasts? If we don't find it again, we may never know.

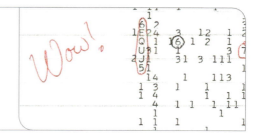

An Alien's Guide to Earth

Would Aliens Come to Save Us or Destroy Us?

n our interactions with other living species, we eradicate them, domesticate them, love them, even worship them. Our behavior is often cultural: People in the U.S. raise cows to eat; people in India protect cows as sacred. So we can't imagine future human-alien interactions based on our current relations with other species alone.

"If aliens came to Earth across the galaxy, it means they've got more advanced technology than we do, so the fear factor makes us think they'll all be evil," Neil explains. "And what I think will save us is that they'll study Earth really hard, and they'll look at the species that calls itself dominant, and they will conclude, based on all the evidence that's available to them, that there is no sign of intelligent life on Earth, and thereby leave us alone." ■

Could Earth be a target for an alien attack?

NEIL: *When was the last time you walked by a pile of worms and said, "Gee, I wonder what they're thinking? I want to make peace with them."*

LYNNE KOPLITZ: *But I've also seen people stomp 'em, just because they were there.*

How Could the Klingons Have Become a Spacefaring Species?

Comedian and *Star Trek* geek Leighann Lord wonders: Given that the Klingon civilization of the *Star Trek* universe is all about combat, conquest, and glory, wouldn't they have destroyed themselves with advanced technology long before they even got off their home planet? Not necessarily, says Neil: "The culture of war is not inconsistent with the culture of technology. In fact, wars drive science. It's a pain to admit that, but it's true. The urge to survive creates extraordinary creative impulses in people to invent something that will make one person survive better than the other, and it's usually in the form of weaponry."

Think about it: We humans have the military technology to destroy our species, but we haven't done it. Maybe during their formative times, the Klingons had political and social restraints on their warlike tendencies as well. Or perhaps the Klingons could have developed their combative ways after they secured their path to the stars. ■

Klingons in Star Trek III: The Search for Spock *(1984)*

"I can imagine an alien species that is [just] energy . . .
The problem is, energy—it's hard to create form out of energy.
When energy becomes matter, you can make molecules and things . . .
Without it, it's amorphous, and it might be harder
to make amorphous life than material life."
—DR. NEIL DEGRASSE TYSON, ASTRO-AMORPHO-CIST

CHAPTER FIVE

When Will We Be Able to Time-Travel?

This is the real frontier of science fiction. Sure, you can change the universe quite a bit with things like black holes and superpowers and star-killer planets. But if you can change the past, you can change *everything* about the universe in the blink of an eye—and nobody would even remember.

Could we go back in time—really manipulate time to that extent, regardless of what technological level we achieve? What exactly is time, anyway? If we think of time as Albert Einstein did—a dimension like length, width, or height, with its own special properties—then the bending of both space and time is in play. Warping and folding space might even allow us to break the fundamental limit of motion in the universe—the speed of light.

One thing is for sure: We can travel forward in time, no problem. We already know the science of how to stay young far into the future while the world ages around us. That's called time dilation—we just don't know how to do it yet. Time will tell.

Someday, time machines may bridge the past, present, and future.

YouTube: Neil deGrasse Tyson on Time:
Doctor Who, Star Trek, or Ray Bradbury?

What Is Time?

"Time is one of the most illusive aspects of physics," says cosmologist Dr. Janna Levin. "We almost imagine it spatially. We almost imagine it like a dimension. But I can't look forward in that dimension the way I can look left. And I can't turn around and look back in that dimension the way I can look right."

Albert Einstein explained in his general theory of relativity that time is a dimension, knit together with the three dimensions of space to create the four-dimensional flexible medium we call space-time. When you move in 4-D, your path is called a world line. Dr. Neil deGrasse Tyson, astro-time-icist, explains: "Other than that, [time] is a term in an equation that allows you to localize something on a world line. And a world line is where you are in space and where you are in time . . . You have never been at a place except for at a time. And you have never existed in a time except at a place. The two go together."

Sounds simple enough—with one important wrinkle. Whereas length, width, and height are bidirectional, time is unidirectional. So when you're measuring a four-dimensional distance in our universe—known as an interval of proper time—the time term has the opposite sign of the length, width, and height terms. That's what makes time travel seem so weird—and why moving backward in time is unphysical. ■

"Astronauts age ever so slightly more slowly than you and I do . . . by being in orbit . . . When you're in a jet airplane flying around the world, your time is passing ever so slightly more slowly."

—BILL NYE THE TIME-DILATION GUY

BIOGRAPHY

ᗜ

Albert Einstein, Person of the 20th Century

Albert Einstein (1879–1955) got through school, earned his doctorate, and got a job, all without excelling but all well enough to give him the time and opportunity to do what he really wanted to do: think about challenging problems in physics. Then in a single "year of miracles," 1905, he described theories for three major scientific mysteries and deduced the formula $\mathcal{E} = mc^2$. Einstein gained worldwide fame when his general theory of relativity was confirmed. Later in life, his contemplations of science, education, the cosmos, and human nature earned him the distinction as *Time* magazine's "Person of the Century."

THINK ON THIS ▶ Are There Places Where Time Passes Differently?

The rate of time passing changes from place to place all the time. The effects are tiny in our daily lives but definitely measurable with nanosecond precision. Faster-moving objects, for example, experience time more slowly than slower-moving objects. Gravity also slows you down; if you and a friend were both, say, hanging out near a black hole, the one closer to it would age more slowly.

Sunlight travels to Earth in roughly eight minutes.

"When I wrote my science-fiction story, it was 'zero-point energy.' But that isn't as jazzy as gravitational waves . . . You get the gravitational wave in front of you to be a little lower than the one behind and you . . . whoosh!"

—DR. BUZZ ALDRIN, ASTRONAUT

Cosmic Queries: Time-Keeping

Does Time Ever Go Backward?

Probably not, although we don't know for sure. According to the mathematics of time dilation, exceeding the speed of light creates travel not in *negative* time, but rather *imaginary* time. "We know that there is this sort of clock," cosmologist Dr. Janna Levin says. "The clock never stops, we never turn around and go back. We can never accidentally go the wrong direction in time. It's always pushing us forward."

▶ WHAT ARE TACHYONS?

According to special relativity, only light can move at the speed of light—in our universe, everything else is slower. Particles that travel faster than light are called tachyons. If they exist, they might hold the key to time travel. As Neil puts it: "If I sent you a tachyon signal, you would get it before I sent it. We don't know if they exist, but it works on paper."

▶ HOW CAN WE PERFORM FASTER-THAN-LIGHT EXPERIMENTS?

Astro-boom-icist Dr. Neil deGrasse Tyson answers: "Light moves fastest in a vacuum. It travels a little slower going through air. It travels a little slower going through water . . . Even slower through glass . . . or diamond. Send a particle through there faster than that speed, [and] that will create . . . a mini light boom."

▶ COULD WE CATCH A GRAVITATIONAL WAVE?

Not to be confused with gravity waves, which happen on lakes and oceans, gravitational waves are ripples in space itself—length, width, and height. They're tiny, though, and only huge events—like exploding stars or colliding black holes—can generate them. So surfing on them might be a little tough. ■

A vessel for cryogenic preservation of tissue

Time Travel at the Movies

Did Walt Disney Get It Right With Cryogenics?

Walt Disney with Mickey Mouse

Urban legend has it that Walt Disney had his body frozen at the moment of his death. (Nope; he was cremated.) But would cryogenics be a way to travel into the future without aging? Human biological tissue is probably too fragile to survive being frozen for very long.

"You freeze the people, and they can wake up when they get to their destination," explains Dr. Phil Plait. "We don't know how to do that. You freeze somebody, they're frozen. That's bad . . . I suppose we can send [Disney's] head to Alpha Centauri and see what happens."

"Bad freezer burn, fellas," adds Leighann Lord.

Nevertheless, the idea of frozen human hibernation has been around a long time and exploited by lots of writers and filmmakers—including, ironically, the Walt Disney Studios. Several of the studio's recent superhero movies feature Captain America, who was accidentally frozen in 1945 and thawed out decades later. ■

Cosmic Queries: Time Travel

There Are No Time Travelers—So Does Time Travel Even Exist?

Fraser Cain, the publisher of Universe Today, once asked Neil: "Doesn't the fact that there are no time travelers now prove that time travel will never be invented in the future?"

Neil's answer: "That's a pretty good argument . . . It might be that your time-travel machine can only take you into the future, and then you don't have these paradoxes of killing your grandmother, and then you're never born."

You could, of course, be exceptionally careful. In the television series *Star Trek: Voyager,* time travelers follow what they call the Temporal Prime Directive, sacrificing everything, including even their own existences, to preserve the time line.

In the short story "Ripples in the Dirac Sea" by NASA astrophysicist and award-winning writer Dr. Geoffrey Landis, a time traveler could not commit the "grandmother paradox," because every time he returned to the present, all the changes he'd caused to the past had disappeared. ◼

Time travel is a theme on The Big Bang Theory.

BACK TO BASICS

Can You Travel Back in Time and Erase Yourself?

"Pick a time where two of your ancestors mated and produced one of your ancestors," proposes astro-pacifist Dr. Neil deGrasse Tyson. "All you have to do is prevent them from mating. You don't have to kill anybody." That works as long as only one time line exists in the universe. On the other hand, such a trip may in fact create an entirely new time line in which you would not exist—but which you also would not have to experience.

THINK ON THIS ▶ Does Time Travel Also Mean Space Travel?

Six months ago, because of its motion around the Sun, Earth was nearly 200 million miles from where it is now. So unless some sort of space-time relativity were preserved for time travel, in the way that a bird in flight doesn't shoot off into deep space, dizzying calculations would be required to land where you want. "You gotta specify location as well as time. Otherwise, you're hosed," Neil warns. "It's quite the ballet."

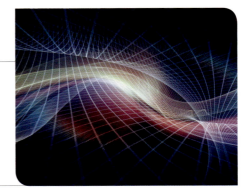

"The idea that we can bend space easier than we can bend ourselves seems, to me, backwards. But that's just the physicist in me. What do I know?"
—DR. CHARLES LIU, ASTROPHYSICIST

The Science of *Interstellar* With Christopher Nolan

Could We Fold Space for Travel?

Space-time is a structure mathematicians call a manifold—which, as its name suggests, could theoretically be folded or at least bent in many directions.

In the general theory of relativity, gravity is the curvature of spacetime; so we'd just need to control a source of gravity large enough to bend space-time inward and then snap it back out again. But gravity travels only at the speed of light—so unless the fold is a permanent crease, the act of folding may not help you achieve faster-than-light travel.

"You could do something like warp drive by contracting space-time between two points, bringing them closer together, jumping across—you don't have to travel 400 light-years—and then you push it back out again. No big deal," says Dr. Janna Levin, cosmologist. ■

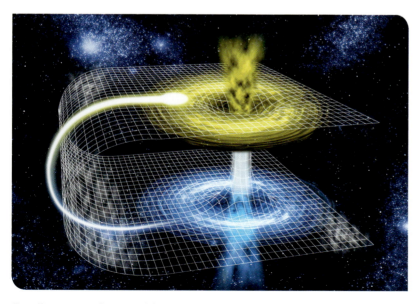

Bending space-time would enable travel between two distant points.

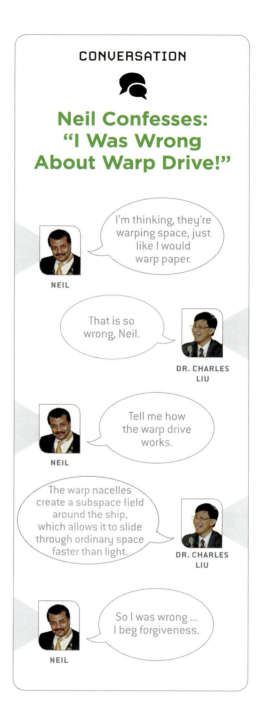

CONVERSATION

Neil Confesses: "I Was Wrong About Warp Drive!"

NEIL: I'm thinking, they're warping space, just like I would warp paper.

DR. CHARLES LIU: That is so wrong, Neil.

NEIL: Tell me how the warp drive works.

DR. CHARLES LIU: The warp nacelles create a subspace field around the ship, which allows it to slide through ordinary space faster than light.

NEIL: So I was wrong ... I beg forgiveness.

"The warp factor is how fast you're able to slide through this subspace stuff."
—DR. CHARLES LIU, ASTROPHYSICIST

Could We Use Black Holes or a TARDIS to Time-Travel?

DID YOU KNOW
If you travel to the black hole Cygnus X-1 in a year, stay a year, and then come back in a year, you will have aged only three years, but traveled 1,201 years into the future.

Black holes are fun mathematical constructs and real physical objects. If a black hole spins fast enough, the geometry of the black hole changes—the "point singularity" at the center becomes a ring, and a doughnut-shaped region forms in which an object can theoretically go at any speed—even exceeding the speed of light—and maybe even time-travel backward!

A totally fictional, but possibly more entertaining, way to time-travel with a black hole is to use it as a power source. In the *Doctor Who* television series, the space-and-time-hopping TARDIS vehicle is powered by the Eye of Harmony—a huge collapsing star frozen in time at the moment it was starting to form a black hole. But the entire amount of energy produced by such a stellar collapse—a Type II supernova—would still not be nearly enough to power everything a TARDIS can do. ∎

"It turns out if you have a rotating black hole, there's a trajectory through it where you could possibly come out the other side before when you left."

—DR. NEIL DEGRASSE TYSON

The *TARDIS* transports the main character in Doctor Who *anywhere in time and space.*

THINK ON THIS ▶ Did *Contact* Get It Right?

In the 1997 movie *Contact*, the protagonist travels in a pod designed to alien specifications. During her journey, 18 hours of time pass while everyone watching her on the ground experiences just a second or two. If her trip had been relativistic—that is, she had been somehow sped up—the time-dilation effects should have been reversed. But hey, this is alien stuff—so who knows what technology they'd have, right?

YouTube: Neil deGrasse Tyson Discusses
Time Travel in *Interstellar*

Is Time Travel Possible Using a Fifth Dimension?

n Madeleine L'Engle's classic novel *A Wrinkle in Time*, the characters traveled rapidly through the fifth dimension. This neat trick allowed them to zoom around the universe, have adventures on strange planets, and still get home in time for dinner. We have no idea if this will ever happen for real, but Dr. Neil deGrasse Tyson, astro-tesser-ist, thinks it certainly could: "If you go to a higher dimension, it is not unrealistic to think that you step out of the time dimension and now you look at time as though we look at space . . . If your whole time line is laid out in front of you, then you have access to it and you can jump in at any point and relive it . . . What does it mean to jump into it and then change something if it's already there?" ∎

The fifth dimension in Interstellar *(2014) is predicated in real physics.*

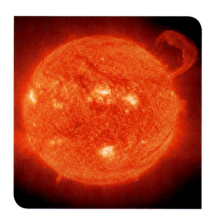

TOUR GUIDE

Can You Fly Into the Sun to Time-Travel?

In both the original *Star Trek* TV series and the movie *Star Trek IV: The Voyage Home,* the crew members of the *Enterprise* hurl themselves backward and forward through time by piloting their ship toward the Sun and using it as a gravitational slingshot. Could that work? "You cannot do what they did in the movie," Neil tells us. "So just nip that one in the bud. The only way you could do something like that is if you had vastly more powerful gravity than the Sun."

"You can ask, 'When was I born?'
Well, you were always born.
'When did I go to college?'
You're always going to college.
'When did I die?'
You're always dying."

—DR. NEIL DEGRASSE TYSON,
ON VIEWING YOUR TIME LINE
FROM A HIGHER DIMENSION

YouTube: Neil deGrasse Tyson on Wormholes and Time Travel

Could Wormholes Be Used to Travel Someday?

Actor George Takei, who played Sulu on *Star Trek,* asked Neil for his thoughts on the ultimate transportation system through space and time. Here's what he had to say: "Wormholes? We're not going anywhere without them . . . but we don't have the command over space-time, matter, and energy yet to make one . . . Yeah, I don't see why [we won't get there eventually] . . . The day we can summon the energy of a galaxy, the mass of all the stars in the supercluster . . . I can imagine manipulating matter, energy—putting curved pockets within the fabric of space between you and your destination—and then the universe becomes a 'wormhole Swiss cheese' set of highways. And then you go wherever you want." ■

"We don't wield enough power over energy in the universe to just create wormholes on a whim. If we harnessed all the energy produced in all the several hundred billion stars of the Milky Way galaxy, that would be enough."

—DR. NEIL DEGRASSE TYSON, ASTRO-WHIM-SICIST

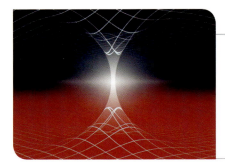

THINK ON THIS ▶ **So When Will This All Come True?**

George Takei asked Neil how many lifetimes it'll be before humans are able to sculpt wormhole highways into space-time. His answer: "I don't know if it's farther away than someone in 1900 saying, 'Oh, we'll never get to the moon,' and then 69 years later we're leaving boot prints." Will we ever know? We do know this for sure: The questions we will have to answer are the ones we have yet to ask.

Acknowledgments

The heart and dedication of each member of *StarTalk*'s production team, past and present, is manifest in every show, and can be felt across all pages of this book.

The producers of *StarTalk* are further grateful to deputy editor Hilary Black at National Geographic Books for conjuring the idea of a *StarTalk* book and for assembling the dream team of associate editor Allyson Dickman, senior editor Susan Tyler Hitchcock, art director Sanaa Akkach, director of photography Susan S. Blair, photo editor Kristin Sladen, production editor Judith Klein, photo assistant Patrick Bagley, editorial assistant Moriah Petty, imaging technician Wendy Smith, associate designer Katie Olsen, and assistant designer Nicole Miller, who together wrangled such diverse content into a stunning visual expression of the show itself.

The soul of *StarTalk* lives within the banter, humor, and content of our shows, offered by our guests and by our studio experts. They're hewn from pop culture, politics, and academia, and we warmly recognize those who were referenced in these pages: Marc Abrahams, James Aguiar, Alan Alda, Buzz Aldrin, Neil Armstrong, David Attenborough, Dan Aykroyd, Hank Azaria, Heather Berlin, Mayim Bialik, Michael Ian Black, Cory Booker, Anthony Bourdain, Charles Bourland, Max Brooks, LeVar Burton, David Byrne, Fraser Cain, Jimmy Carter, Wyatt Cenac, Andrew Chaikin, Michael Che, Bill Clinton, Stephen Colbert, David Cope, Brian Cox, David Crosby, R. Walter Cunningham, Richard Dawkins, Peter Diamandis, His Holiness Gyalwang Drukpa, Ann Druyan, Christopher Emdin, Carter Emmart, Laurence Fishburne, Helen Fisher, Andrew Freedman, Jim Gaffigan, Laurie Garrett, Malcolm Gladwell, John Glenn, Stephen Gorevan, Temple Grandin, David Grinspoon, Josh Groban, GZA, Chris Hadfield, Maeve Higgins, John Hodgman, G. Scott Hubbard, Arianna Huffington, Jamie Hyneman, Penn Jillette, James Kakalios, Michio Kaku, Stephen Keil, Robert F. Kennedy, Jr., Lynne Koplitz, Mark Kurlansky, Roger Launias, Sally Le Page, Judith Lean, Janna Levin, Ian Lipkin, John Logsdon, Seth MacFarlane, Sandra Hall Magnus, Bill Maher, Amy Mainzer, the Rev. James R. Martin, Susana Martinez-Conde, Mike Massimino, Peter Max, James McGaha, Seth Meyers, Moby, Elon Musk, Marion Nestle, Nichelle Nichols, Miles O'Brien, John Oliver, Rebecca Oppenheimer, Yvonne Pendleton, Phil Plait, Carolyn Porco, Ainissa Ramirez, Alan Rickman, David Rind, Joan Rivers, Mary Roach, Joe Rogan, Cynthia Rosenzweig, Paul Rudd, Tess Russo, Jeffrey Ryan, Oliver Sacks, Cara Santa Maria, Susan Sarandon, Adam Savage, Dan Savage, Kristen Schaal, Andy Serkis, Seth Shostak, Mark Siddall, Jason Silva, Lee Silver, Jennifer Simonetti-Bryan, Steven Soter, Brent Spiner, Steven Squyres, Melissa Sterry, Jon Stewart, Biz Stone, Nina Strohminger, Jason Sudeikis, George Takei, Ian Tattersall, Travis Taylor, Clive Thompson, The Tweet of God, Shannon Walker, James Webster, Andy Weir, Ruth Westheimer, Wil Wheaton, Peter Whiteley, Larry Wilmore, Will Wright.

Neil deGrasse Tyson, Chuck Nice, and Jeff Jarvis on the set of StarTalk

StarTalk is filmed in front of a live audience in the American Museum of Natural History's Cullman Hall of the Universe in the Rose Center for the Earth and Space. Host Neil deGrasse Tyson is the Frederick P. Rose Director of the Hayden Planetarium.

Contributors

Neil deGrasse Tyson, *StarTalk* host, is a graduate of the prestigious Bronx High School of Science. He earned his B.A. in physics from Harvard and his Ph.D. in astrophysics from Columbia. He is the author of dozens of research papers and ten books—two of which became PBS *NOVA* specials. Neil is the recipient of 19 honorary doctoral degrees, and is host of *Cosmos: A SpaceTime Odyssey,* a 13-part series on the universe that aired on Fox and the National Geographic Channel. His late-night *StarTalk* program is the first television talk show on science, now appearing on the National Geographic Channel. He lives in New York City with his wife and two children.

Charles Liu, author, is an astrophysics professor at the City University of New York's College of Staten Island and an associate with the Hayden Planetarium and Department of Astrophysics at the American Museum of Natural History in New York. His research focuses on colliding galaxies, quasars, and the star formation history of the universe. He earned degrees from Harvard and the University of Arizona, and held postdoctoral positions at Kitt Peak National Observatory and Columbia. Together with co-authors Robert Irion and Dr. Neil deGrasse Tyson, he received the 2001 American Institute of Physics Science Writing Award for the book *One Universe: At Home in the Cosmos.* He is also the author of *The Handy Astronomy Answer Book,* now in its third edition. He and his wife, who is way smarter than he is, have three children, who are also way smarter than he is.

Jeffrey Lee Simons, editor, is the social media director of *StarTalk Radio.* When he's not busy engaging with the greatest fans in the universe, he's also a writer. He is the author of *Spirit in Realtime,* the first installment of his near-future virtual reality/cyberpunk young adult science-fiction series. Simons is also the co-author of the cause-related marketing handbook *Making Money While Making a Difference,* with Dr. Richard Steckel, and the author of two works with noted artist Viktor Koen: *Lexicon: Words and Images of*

Strange and *Toyphabet.* He has a B.A. in literature from Georgetown University. He lives in New Jersey with his wife, his daughter, and four cats.

Additional Contributors

Bill Nye, *StarTalk* host, holds a B.S. in mechanical engineering from Cornell University and worked as an engineer at Boeing before creating and hosting his Emmy award–winning PBS/Discovery Channel show *Bill Nye the Science Guy.*

Eugene Mirman, *StarTalk* cohost, is a comedian and actor. He voices Gene on Fox's Emmy award–winning animated series *Bob's Burgers* and has had two Comedy Central specials and five comedy albums. He lives in Brooklyn, New York.

Chuck Nice, *StarTalk* cohost, is an 18-year veteran of stand-up comedy, with a rich history in television and on radio. He is the host of *Buy Like a Mega Millionaire* on HGTV, *The Juice* on Veria Living, and *The Hot 10* on Centric.

Leighann Lord, *StarTalk* cohost, is a stand-up comedian named "The Most Thought Provoking Black Female Comic" by NYC Black Comedy. She is also a contributor to the Huffington Post and the author of several books, including *Leighann Lord's Dict Jokes,* volumes one and two.

Illustrations Credits

FRONT COVER: (Neil deGrasse Tyson), Miller Mobley/AUGUST; (space behind Tyson), NASA, ESA and A. Nota (STScI/ESA); (crystal skull), Jamen Percy/Shutterstock; (cocktail), James Brey/Getty Images; (Saturn & satellites), John Foxx/Getty Images; (shuttle), NASA; (microphone), rangizzz/Shutterstock; (microscope), Vitaly Korovin/Shutterstock; (stethoscope), Edward Westmacott/Shutterstock; (Hubble telescope), NASA; (asteroids), Scott Tysick/Getty Images; (background), European Space Agency, NASA and Felix Mirabel (the French Atomic Energy Commission & the Institute for Astronomy and Space Physics/Conicet of Argentina); BACK COVER: (UP LE), SVF2/Getty Images; (UP RT), Eric Isselee/Shutterstock; (LO LE), Dmitry/Shutterstock; (LO RT), Giovanni Cancemi/Shutterstock; 1, NASA/Corbis; 2-3, NASA/JPL-Caltech; 6, Fedorov Oleksiy/Shutterstock; 7, FOX via Getty Images; 8 (UP), Courtesy National Geographic Channel; 8 (LO), John A. Davis/Shutterstock; 9 (UP), NASA/Bill Ingalls; 9 (LO), JoeFotoSS/Shutterstock; 12-13, NASA, ESA, and the Hubble Heritage Team (STScI/AURA)—ESA/Hubble Collaboration; 15, NASA/JPL-Caltech; 16, Dmitry/Shutterstock; 17 (UP), YURI KOCHETKOV/epa/Corbis; 17 (LO), Pilate/Shutterstock; 18, NASA; 19 (UP LE), NASA; 19 (UP CTR), National Air and Space Museum, Smithsonian Institution; 19 (UP RT), NASA; 19 (LO), Final Frontier Design, photo by Virgil Calejesan; 20 (BOTH), Photo Still(s) from Star Trek—Courtesy of CBS Television Studios; 21 (UP), Roger Harris/Science Source; 21 (CTR), Dudarev Mikhail/Shutterstock; 21 (LO), NASA; 22 (UP), NASA; 22 (LO), RIA Novosti/Science Source; 22-3, pixelparticle/Shutterstock; 23 (UP), NASA; 23 (CTR), Li junfeng—Imagine china via AP Images; 23 (LO), NASA/Corbis; 24 (UP), Sebastien Micke/Paris Match via Getty Images; 24 (LO), Galen Rowell/Corbis; 25 (LE), *Packing for Mars* by Mary Roach, W.W. Norton & Company, 2011; 25 (RT-ALL), NASA; 26, FOX via Getty Images; 27 (UP), James O. Davies; 27 (LO), NASA/JPL/MSSS; 28, NASA; 29 (UP), Fabio Berti/Shutterstock; 29 (LO), NASA; 30 (BOTH), NASA; 31 (UP RT), AFP/Getty Images; 31 (OTHERS), NASA; 33, Elena Shashkina/Shutterstock; 34 (UP), bonchan/Shutterstock; 34 (LO), www.Billion Photos.com/Shutterstock; 35 (BOTH), NASA; 36 (UP), Tarasyuk Igor/Shutterstock; 36 (LO), Joy Tasa/Shutterstock; 37 (UP), Courtesy of StarTalk Radio; 37 (CTR LE), Sunny Forest/Shutterstock; 37 (CTR), margouillat photo/Shutterstock; 37 (CTR RT), photocat5/Shutterstock; 37 (LO), NBC NewsWire/Peter Kramer/Nov. 23, 2011, via Getty Images; 38 (UP LE), NASA; 38 (UP RT), National Air and Space Museum, Smithsonian Institution; 38 (CTR LE), National Air and Space Museum, Smithsonian Institution; 38 (CTR RT), SSPL/Getty Images; 38 (LO LE), NASA/Roger Ressmeyer/Corbis/VCG/Getty Images; 38 (LO RT), Stringer/AFP/Getty Images; 39 (UP LE), NASA; 39 (UP RT), molotok743/Shutterstock; 39 (CTR), NASA; 39 (LO LE), NASA/Gioia Massa; 39 (LO RT), Keith Homan/Shutterstock.com; 40 (UP), NASA/Terry W. Virts; 40 (LO LE), Lukas Gojda/Shutterstock; 40 (LO RT), Valentyn Volkov/Shutterstock; 41 (UP), Detlev van Ravenswaay/SPL/Getty Images; 41 (LO), Vadim Sadovski/Shutterstock; 42 (UP), Stephen Coburn/Shutterstock; 42 (LO), NASA/JPL-Caltech/Univ. of Arizona; 43 (UP LE), Philippe Desnerck/Getty Images; 43 (LO LE), Bill O'Leary/The Washington Post/Getty Images; 43 (RT-MINTS), courtesy Wrigley; 43 (RT-CANDY BAR), bestv/Shutterstock; 43 (RT-MOON PIE), Chattanooga Bakery, Inc.; 43 (RT-SODA), iStock.com/jfmdesign; 44 (UP), Edward Kinsman/Science Source/Getty Images; 44 (LO), Hummy/Shutterstock; 45 (UP), Olga Nayashkova/Shutterstock; 45 (LO), Igor Stevanovic/Shutterstock; 47, Victor Habbick/Visuals Unlimited, Inc./Getty Images; 48 (UP), asharkyu/Shutterstock; 48 (LO), Maxim Garagulin/Shutterstock; 49 (UP), Cosmos Studios/Fuzzy Door Productions/National Geographic/Six Point Harness/The Kobal Collection; 49 (LO), Lilly Lawrence/Getty Images; 50 (UP), LAGUNA DESIGN/Getty Images; 50 (LO), LungLee/Shutterstock.com; 51 (UP), Peter Ginter/Getty Images; 51 (LO), ESA & Planck Collaboration/Rosat/Digitised Sky Survey; 52 (UP), C.P Storm/flickr (https://creativecommons.org/licenses/by/2.0/legalcode); 52 (LO), Mark Garlick/Science Photo Library/Getty Images; 53 (UP),

Alain r/Wikimedia Commons (https://creativecommons.org/licenses/by-sa/2.5/legalcode); 53 (LO), Roman Sigaev/Shutterstock; 54 (UP), NASA, H.E. Bond and E. Nelan (Space Telescope Science Institute, Baltimore, Md.), M. Barstow and M. Burleigh (University of Leicester, U.K.), and J.B. Holberg (University of Arizona); 54 (LO-XRAY); NASA/CXC/ASU/J. Hester et al.; 54 (LO-OPTICAL); NASA/HST/ASU/J. Hester et al.; 54 (LO-RADIO); VLA/NRAO; 55 (UP), Mark Garlick/Science Photo Library/Getty Images; 55 (CTR), Ian Cuming/Getty Images; 55 (LO), NASA, ESA, D. Coe (NASA Jet Propulsion Laboratory/California Institute of Technology, and Space Telescope Science Institute), N. Benítez (Institute of Astrophysics of Andalucía, Spain), T. Broadhurst (University of the Basque Country, Spain), and H. Ford (Johns Hopkins University, USA) (http://creativecommons.org/licenses/by/3.0/legalcode); 56, Chiung-hung Huang/Alamy Stock Photo; 57 (UP), Science Source/Getty Images; 57 (LO), ESA/Hubble & NASA (http://creativecommons.org/licenses/by/3.0/legalcode); 58 (UP), Igor Zh./Shutterstock; 58 (LO), wanphen charwarung/Shutterstock; 59 (UP), Cultura RM Exclusive/Liam Norris/Getty Images; 59 (CTR), Gary Brown/Science Source; 59 (LO), Tim Graham/Getty Images; 61, NASA; 62 (UP), NASA/JPL-Caltech; 62 (UP-PORTRAIT), Frederick M. Brown/Getty Images; 62 (LO), Josh Spradling/The Planetary Society; 62 (LO-PORTRAIT), Larry Busacca/Getty Images; 63 (UP), NASA/JPL-Caltech; 63 (UP-PORTRAIT), Cornell University; 63 (LO LE), NASA; 63 (LO LE-PORTRAIT), Phil Mumford; 63 (LO RT), NASA/JPL-Caltech/MSSS; 63 (LO RT-PORTRAIT), Alan Fischer; 64 (LE), NASA/Marshall Space Flight Center; 64 (RT), NASA; 65 (UP), Amblin/Dreamworks/WB/The Kobal Collection/James, David; 65 (LO), Nicescene/Shutterstock.com; 66 (UP), Courtesy of StarTalk Radio; 66 (CTR), querbeet/Getty Images; 66 (LO), David Paul Morris/Getty Images; 67 (UP), Jason Kempin/Getty Images; 67 (LO), Victor Habbick/Shutterstock; 68 (UP), World History Archive/Alamy Stock Photo; 68 (LO), Rick Kern/Getty Images; 69 (UP), Marina Sun/Shutterstock; 69 (LO), AP Images/Bruce Weaver; 70 (UP), Charles Taylor/Shutterstock; 70 (LO), ifong/Shutterstock; 71 (UP), NASA/Ames/JPL-Caltech; 71 (LO LE), Photobank gallery/Shutterstock; 71 (LO RT), shiva3d/Shutterstock; 72, solarseven/Shutterstock; 73 (UP), Tappasan Phurisamrit/Shutterstock; 73 (LO), Joshua Sharp/Shutterstock; 75, NASA/JPL-Caltech; 76 (PLUTO), NASA/JHUAPL/SwRI; 76-7, NASA/Ames/JPL-Caltech; 78 (UP), NASA/JPL/Space Science Institute; 78 (CTR), Don Davis/NASA; 78 (LO), Courtesy of http://tabbythis.com; 79, John White Photos/Alamy Stock Photo; 80 (UP), NASA/JPL-Caltech; 80 (LO), ESA 2010 MPS for OSIRIS Team MPS/UPD/LAM/IAA/RSSD/INTA/UPM/DASP/IDA; 81 (UP), NASA/JPL-Caltech/T. Pyle (SSC); 81 (LO), Alessio Botticelli/FilmMagic/Getty Images; 82 (UP LE), Gary Ombler/Getty Images; 82 (UP RT), NASA/JHUAPL/SwRI; 82 (CTR), NASA/JPL-Caltech/UCLA/MPS/DLR/IDA; 82 (LO), Thomas Heaton/Getty Images; 83 (UP), NASA/JPL-Caltech; 83 (CTR), NASA/JPL; 83 (LO), Dennis di Cicco/Corbis; 84, J Foster/Getty Images; 85 (UP), NASA/JPL-Caltech/UCLA; 85 (LO), zhangyouyang/Shutterstock; 86 (UP), fortuna777/Shutterstock; 86 (LO), inhauscreative/Getty Images; 87 (UP), igor.stevanovic/Shutterstock; 87 (CTR), G10ck/Shutterstock; 87 (LO), givaga/Shutterstock; 88, Courtesy of Lockheed Martin; 89 (UP), Arthimedes/Shutterstock; 89 (LO), NASA/NOAA/GSFC/Suomi NPP/VIIRS/Norman Kuring; 90-91, ESA/NASA; 93, wragg/Getty Images; 94 (UP LE), NASA; 94 (UP RT), Earth Science and Remote Sensing Unit, NASA Johnson Space Center (http://eol.jsc.nasa.gov/); 94 (LO), NASA/JPL-Caltech/SSI; 95 (UP), NASA; 95 (CTR), NASA/JPL; 95 (LO), Scott Kelly/NASA; 96, NASA; 97 (UP), Mickey Adair/Michael Ochs Archives/Getty Images; 97 (LO), NASA/JPL-Caltech/SSI; 98 (UP), Detlev van Ravenswaay/Science Source; 98 (LO), David A. Hardy/Science Source; 99, David Lees/Corbis; 100 (UP), Matt Cardy/Getty Images; 100 (LO), Claudio Divizia/Shutterstock; 101 (UP), James D. Balog; 101 (LO), bikeriderlondon/Shutterstock; 102 (UP), Henning Dalhoff/SPL/Getty Images; 102 (LO), Kichigin/Shutterstock; 103 (UP), aodaodaodaod/

Since 1888, the National Geographic Society has funded more than 12,000 research, exploration, and preservation projects around the world. National Geographic Partners distributes a portion of the funds it receives from your purchase to National Geographic Society to support programs including the conservation of animals and their habitats.

National Geographic Partners
1145 17th Street NW
Washington, DC 20036-4688 USA

Become a member of National Geographic and activate your benefits today at natgeo.com/jointoday.

For information about special discounts for bulk purchases, please contact National Geographic Books Special Sales: specialsales@natgeo.com

For rights or permissions inquiries, please contact National Geographic Books Subsidiary Rights: bookrights@natgeo.com

ISBN: 978-1-4262-1727-2

Printed in the United States of America

16/RRDK-RRDML/1

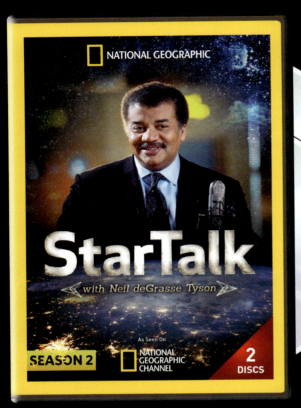 
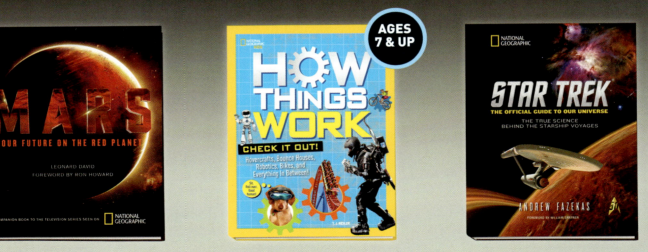
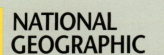